LabVIEW™

A Developer's Guide
to Real World Integration

Edited by

Ian Fairweather
Anne Brumfield

CRC Press
Taylor & Francis Group
Boca Raton London New York

CRC Press is an imprint of the
Taylor & Francis Group, an **informa** business

A CHAPMAN & HALL BOOK

CRC Press
Taylor & Francis Group
6000 Broken Sound Parkway NW, Suite 300
Boca Raton, FL 33487-2742

© 2012 by Taylor & Francis Group, LLC
CRC Press is an imprint of Taylor & Francis Group, an Informa business

No claim to original U.S. Government works

Printed in the United States of America on acid-free paper
Version Date: 20111014

International Standard Book Number: 978-1-4398-3981-2 (Hardback)

Library of Congress Cataloging-in-Publication Data

LabVIEW : a developer's guide to real world integration / [edited by] Ian Fairweather and Anne Brumfield.
   p. cm.
   Includes bibliographical references and index.
   ISBN 978-1-4398-3981-2 (hardback : acid-free paper)
   1. LabVIEW. 2. Computer graphics. 3. System integration. I. Fairweather, Ian. II. Brumfield, Anne.

T385.L3325 2011
006.6'63--dc23   2011027433

Visit the Taylor & Francis Web site at
http://www.taylorandfrancis.com

and the CRC Press Web site at
http://www.crcpress.com

Table of Contents

About This Book

LabVIEW™ has been in existence now for several decades and during that time has emerged to be one of the preeminent platforms for the development of data acquisition and data analysis programs. National Instruments Corporation has continually developed and improved its product and has been very active in ensuring that continuing feedback from users is acted on and implemented in the form of new functionality, toolkits, and upgrades.

Many texts have already been written describing how to use LabVIEW. This book is not another "hello world" description of how to operate or use basic LabVIEW commands, nor is it a highbrow text describing its more intricate and complex functionality. This book is intended to be a down-to-earth guide on how LabVIEW can be integrated into practical real-world applications. It is aimed at LabVIEW users who are already involved in developing stand-alone applications and who are competent developers possessing reasonable knowledge of PC operating systems.

This text is a gathering of diverse chapters written by experienced and practical LabVIEW developers and engineers who have utilized LabVIEW's power and simplicity to great advantage in their applications. Many of the chapters provide information on exciting new technologies and how these can be implemented in LabVIEW to provide novel solutions to new or existing problems. The text describes how LabVIEW has been pivotal in solving real-world challenges. Novel tricks and tips for integrating LabVIEW with third-party hardware and software provide the reader with new approaches and techniques.

Chapters in this book are presented as complete within themselves and are not contingent upon readers having to read sequentially through the text. Projects and Virtual Instruments (VIs) are provided for most of the programs and utilities described. The software described in the text has been developed on PC-based platforms, and many applications use free downloadable third-party software or drivers. The CD that accompanies this book contains various authors' software contributions.

The editors wish to sincerely thank all of the contributing authors for their generosity in providing the book's chapters. Without their efforts and experience this book would not have been possible. Thanks are also extended to the publishers and their staff who have provided patient and professional guidance during the development processes.

About the Authors

Steffan Benamou is founder of Aledyne Engineering, a software and hardware design consulting firm located in the Bay Area, California. Steffan is recognized by National Instruments as a certified developer, and his areas of expertise include embedded systems, user interface design, data acquisition, and hardware connectivity. He has a background in electrical engineering and software development and received his bachelor's in electrical engineering from California Polytechnic State–San Luis Obispo and master's, also in electrical engineering with an emphasis in analog integrated circuit design, from Santa Clara University. Benamou has over 8 years of experience working for a prominent medical device manufacturer and has held many positions from test engineer to design engineer and lead software architect on multiple projects. During his tenure, he has developed many LabVIEW-based test systems as well as numerous embedded designs for Class II medical devices.

Anne M. Brumfield is currently an independent consultant in sports testing and medical device development. Her research areas include sensor integration, wireless data acquisition, and signal and image processing. Brumfield worked in cancer and diabetes research at the University of Pittsburgh for over a decade. She has worked as a biomedical engineer at the National Institute for Occupational Safety and Health in the Engineering & Control Technology Branch for several years and has publications in the application of sensors in human monitoring. She is currently a member of IEEE and the IEEE Robotics and Automation Society.

Nesimi Ertugrul received his bachelor's (1985) and master's (1989) in electrical and electronic and communication engineering and his Ph.D. from the University of Newcastle upon Tyne, United Kingdom, in 1993. He has been with the University of Adelaide since 1994, where he is associate professor. Ertugrul has published two books, two book chapters, and a number of journal articles and conference publications. He is a member of IEEE and serves on the advisory board of the *International Journal of Engineering Education (IJEE)*. His primary research interests are sensorless operation of switched motors, fault-tolerant motor drives, condition monitoring, and electric vehicles. Ertugrul is the author of *LabVIEW for Electric Circuits, Machines, Drives and Laboratories* and also the guest editor in two special issues of *IJEE* on LabVIEW.

Ian Fairweather is currently on staff at Victoria University's (VU's) School of Sport and Exercise Science (SES) as an engineer principally involved with software and hardware developments revolving around sports science and sports engineering. Fairweather's background is in electronics and applied science, and he has worked for many years in the instrumentation and data acquisition end of elite athlete performance measurement and optimization. He has also previously worked in VU's Department of Applied Physics as research fellow for the Centers for Disease Control and Prevention (CDC), National Institute for Occupational Safety and Health (NIOSH), and more recently as a contract software developer for several U.S.-based sports and medical equipment startup companies. He is the current secretary of the Health & Exercise Science Technologists Association (HESTA), a member of the IEEE, and associate researcher with the Institute for Sport, Exercise and Active Living (ISEAL). Fairweather also provides ongoing consultancy services to the wider sports industry including the Australian and Victorian institutes of sport as well as other academic and research institutions and is currently involved in the establishment of undergraduate and postgraduate programs in sports engineering at Victoria University.

Vanishree Gopalakrishna is currently a Ph.D. candidate in the Department of Electrical Engineering at the University of Texas at Dallas. Her research interests include real-time speech processing for cochlear implants and pattern recognition.

Philipos Loizou is a professor and holds the Cecil and Ida Green Chair in the Department of Electrical Engineering at the University of Texas at Dallas. He is a Fellow of the Acoustical Society of America. His research interests include signal processing, speech processing, and cochlear implants.

Nasser Kehtarnavaz is professor of electrical engineering and director of the Signal and Image Processing Laboratory at the University of Texas at Dallas. He has nearly 25 years of academic and industrial experience in areas including signal and image processing, real-time signal and image processing, and biomedical image analysis. He has authored or coauthored 8 books and more than 200 papers in these areas. He is currently chair of the Dallas chapter of the IEEE Signal Processing Society and coeditor-in-chief of the *Journal of Real-Time Image Processing*.

Brian MacCleery is senior product manager for Energy Efficiency and Renewable Energy at National Instruments (NI) in Austin, Texas. He manages market development operations for NI green engineering products worldwide. MacCleery helps small businesses and entrepreneurs bring their green ideas to market using state-of-the-art graphical system design technology from National Instruments. MacCleery holds electrical engineering degrees from Virginia Tech where he completed his master's in electromagnetic propulsion and power electronics instrumentation, modeling, and control.

Arturo Molina is rector and vice president for Research and Development at the Instituto Tecnológico de Monterrey. Previously, he was director at Instituto Tecnológico de Monterrey at the Mexico City Campus. Molina holds computer science engineering degrees from ITESM where he completed his master's in computer science. In 1992 he received his Ph.D. in Mechanical Engineering from the Technical University of Budapest, Hungary. In 1995, he earned a Ph.D. in manufacturing systems from the Loughborough University of Technology in the UK.

Craig Moore is systems integrator and software developer at the product development company Bjorksten | bit7 in Madison, Wisconsin. Moore has used LabVIEW and National Instruments' products for over 20 years to solve problems in areas such as scientific instrument development, automated test and measurement, image processing, data analysis, and consumer product development. He is also an NIWeek 2004 Semiconductor Category Winner for Best Application of Virtual Instrumentation.

Hiram Ponce received the M.Sc. degree in engineering, specializing on intelligent control systems from the ITESM-CCM University in Mexico City in 2010. He recently earned a Ph.D. in computer science focusing on developing new artificial intelligence bio-inspired techniques in organic chemistry. He is an assistant researcher at the ITESM-CCM University in Mexico City. His primary areas of interests are intelligent control systems and artificial intelligence.

Pedro Ponce received his bachelor's in automation and control systems and his master's and Ph.D. in electrical engineering with a focus on control systems. His areas of interest are electric machines, power electronics, robotics, artificial intelligence, control systems, and renewable energy. He has worked as field and design engineer in speed control and as industrial developer project engineer for 9 years. He has 15 patents, 4 books, and more than 60 research publications. He is director of the Ph.D. and master's programs in engineering at Instituto Tecnologico de Monterrey, Mexico City.

Bill VanArsdale currently works for Boeing's NAVSTAR GPS Operations at Schriever Air Force Base near Colorado Springs, Colorado. Previously, he consulted through his company SmartMethods LLC, and spent over 2 decades as a professor of mechanical engineering at the University of Houston. He has used LabVIEW since its initial release on Macintosh computers. He received his Ph.D. from Cornell University in 1981 in theoretical and applied mechanics.

Chapters in Brief

CHAPTER 1

Excel and ActiveX Automation Using LabVIEW, by *Steffan Benamou*

LabVIEW provides full connectivity capabilities to ActiveX, and many exciting applications can be conceived within the development environment. ActiveX is an expansive topic, but this chapter focuses specifically on the dynamics of the automation of Excel from LabVIEW and provides a simple mechanism that enables the full capabilities of Excel from within the LabVIEW programming environment.

CHAPTER 2

Interacting with Windows Applications, by *Bill VanArsdale*

LabVIEW programs often use other applications to accomplish tasks not readily handled within the development environment. This interaction typically involves a command line interface or properties and methods associated with ActiveX and .NET objects in Windows. This chapter illustrates these techniques by providing solutions to some common application tasks.

CHAPTER 3

A General Architecture for Image Processing, by *Craig Moore*

The chapter describes a framework for a general image-based particle analysis and characterization application known as the Bjorksten Particle Analysis System. Architecture and tools are employed to work toward solutions to many image analysis problems. Special attention is paid to subsets of digital image processing commonly known as particle analysis and edge detection. By defining the boundaries of these particles, they can then be readily characterized and exploited. A worked ready-made solution to these and many other image analysis applications is provided as a building block or tool for inclusion in more complete systems and solutions.

CHAPTER 4

Radio Frequency Identification Read/Write in
Sports Science, by *Ian Fairweather*

This chapter presents complete work solutions and VIs to both the reading and writing of data to and from radio frequency identification (RFID) tags. The chapter focuses on a sports science application using RFID; however, the basic principles and VIs included could be readily adapted to any application requiring their use or implementation. The chapter provides solutions primarily for ISO14443 and ISO15693 protocol RFID tags operating with cheap and readily available reader/writer devices.

CHAPTER 5

Pachube: Sensing the World through the
Internet, by *Anne M. Brumfield*

The Pachube web service is introduced as a novel platform for collaborative sensor monitoring. An overview for the new user is presented, and methods for sending, receiving, and displaying data using the Pachube application programming interface (API) are demonstrated. Specifically, the communication between the Pachube network and LabVIEW using the .NET framework's WebClient class for uploading and downloading via Hypertext Transfer Protocol (HTTP) are discussed.

CHAPTER 6

Power System Applications in LabVIEW, by *Nesimi Ertugrul*

Power systems cover a wide range of diverse applications such as rotating electrical machines, renewable energy systems, power electronics and distribution systems, and display distinct electrical behaviors. This chapter's primary objective is to help the reader to understand the characteristics of a given power system. Another goal is to provide a better interpretation of programs and associated data and results while improving productivity and offering common approaches for future applications and developments.

CHAPTER 7

Recursive Computation of Discrete Wavelet Transform, by *Nasser Kehtarnavaz, Vanishree Gopalakrishna, and Philipos Loizou*

Wavelet transform is increasingly being used in place of short-time Fourier transform in various signal processing applications such as time-frequency analysis, data compression, denoising, and signal classification. Many signal processing applications require wavelet transform to be computed in real time over a moving window. This chapter presents a recursive way of computing discrete wavelet transform (DWT) over a moving window, which has shown to be computationally more efficient than the conventional nonrecursive approach.

CHAPTER 8

Solar Energy, by *Pedro Ponce, Brian MacCleery, Hiram Ponce, and Arturo Molina*

This chapter introduces photovoltaic cells and solar tracking systems focusing and their design, testing, and performance. Photovoltaic cell modeling using LabVIEW is introduced as simple and complex mathematical equations and intelligent model techniques. Solar tracking systems controlled by LabVIEW are presented as proportional–integral–derivative (PID) control techniques and genetic algorithms. Finally, solar tracking systems analysis is discussed taking into account the mechanical modeling of tracker drive motors.

Contributing Authors

Steffan Benamou
Aledyne Engineering
Morgan Hill, California
steffan@aledyne.com

Anne M. Brumfield
Daqtronics, Inc.
Cortland, Ohio
anne.brumfield@gmail.com

Nesimi Ertugrul
School of Electrical and Electronic
    Engineering
The University of Adelaide
Adelaide, Australia
nesimi@eleceng.adelaide.edu.au

Ian Fairweather
School of Sport and Exercise
    Science
Victoria University
Melbourne, Australia
ian.fairweather@vu.edu.au

Vanishree Gopalakrishna
Department of Electrical
    Engineering
University of Texas at Dallas
Richardson, Texas
vani@utdallas.edu

Nasser Kehtarnavaz
Department of Electrical
    Engineering
University of Texas at Dallas
Richardson, Texas
kehtar@utdallas.edu

Philipos Loizou
Department of Electrical
    Engineering
University of Texas at Dallas
Richardson, Texas
loizou@utdallas.edu

Brian MacCleery
National Instruments
Area Renewables & Environment
Austin, Texas
brian.maccleery@ni.com

Arturo Molina
Research and Development,
Instituto Tecnológico de
    Monterrey
Mexico City, Mexico
armolinagtz@itesm.mx

Craig Moore
Bjorksten/bit 7
Madison, Wisconsin
cmoore@bjorksten.com

Hiram Ponce
School of Engineering
Instituto Tecnológico de
    Monterrey
Mexico City, Mexico
hiram.ponce@itesm.mx

Pedro Ponce
School of Engineering
Instituto Tecnológico de
    Monterrey
Mexico City, Mexico
pedro.ponce@itesm.mx

Bill VanArsdalé
The Boeing Company
Colorado Springs, Colorado
william.e.vanarsdale@boeing.com

CHAPTER 1

Excel ActiveX Automation Using LabVIEW

Steffan Benamou

CONTENTS

INTRODUCTION

For the Windows-based LabVIEW users out there, ActiveX provides a means of communication among different applications. Specifically, programs such as Microsoft (MS) Excel, Word, and PowerPoint provide an ActiveX interface that allows them to be controlled from another application. ActiveX can also be used to embed user interface objects from one application into another. LabVIEW provides full connectivity capabilities to ActiveX, and there are many exciting applications that can be conceived within the development environment. Since ActiveX is an expansive topic and cannot be entirely covered here, this chapter will focus specifically on the dynamics of automation of Excel from LabVIEW.

BACKGROUND

In today's engineering world, data acquisition systems are becoming more and more complex and have the ability to acquire much more data at higher sample rates than ever before. However, these datasets are useless unless they can be represented visually and presented to the intended audience in a readable and consistent format. LabVIEW provides a powerful development environment to create dynamic test systems that can measure many complex signals and can analyze the captured data efficiently. When it comes to reporting those results, there are also many ways to do this within LabVIEW. The results can be displayed on the front panel and the data can be analyzed from the user interface. Tables and graphs can be exported to a Hypter Text Markup Language (HTML) or Portable Document Format (PDF) report. But what if the audience would like the results in a format that can be easily viewed and modified from their own machine and within another application? Or what if the data have been acquired by another test system outside of LabVIEW? This is where ActiveX can be very useful and efficient at creating customized Excel reports without the engineer or tester having to even open Excel.

LabVIEW includes a Report Generation Toolkit for Microsoft Office in the Developer's Suite Package that is intended to automate Microsoft Excel and Word from LabVIEW. However, there are certain limitations to this package. The implementation of this package uses a Microsoft graph Object Linking and Embedding (OLE) object for graphing in Excel instead of the Excel built-in Microsoft Excel Chart object. OLE graph objects have a limitation of 4,000 data points as well as limitations of other complex charting features, such as shadowing and 3-D effects offered by Excel 2007. Also, data that is already present in an Excel file cannot be charted without first transferring it to LabVIEW and then calling the graph OLE object to export the data back to Excel. All the features of the Excel chart object can be accessed through the ActiveX interface and LabVIEW without the Report Generation toolkit. Furthermore, with the Excel chart object, existing spreadsheet data can be added to a graph without having to transfer any data to or from the LabVIEW development environment.

CHALLENGE

In this example we are challenged with the task of converting hundreds of tab-delimited files or Excel workbooks containing raw data to professional-looking graphical reports in Microsoft Excel 2007 automated from LabVIEW. This shall be accomplished without using any additional toolkits in LabVIEW. Let's assume another team has a system in place that records raw data from multiple temperature sensors to a different tab-delimited or Excel workbook each day. Temperature data is acquired from these sensors once every 10 seconds; therefore, 8,640 samples exist per sensor per day. A sample file is shown in Figure 1.1. It has been requested that all prerecorded temperature data for each sensor

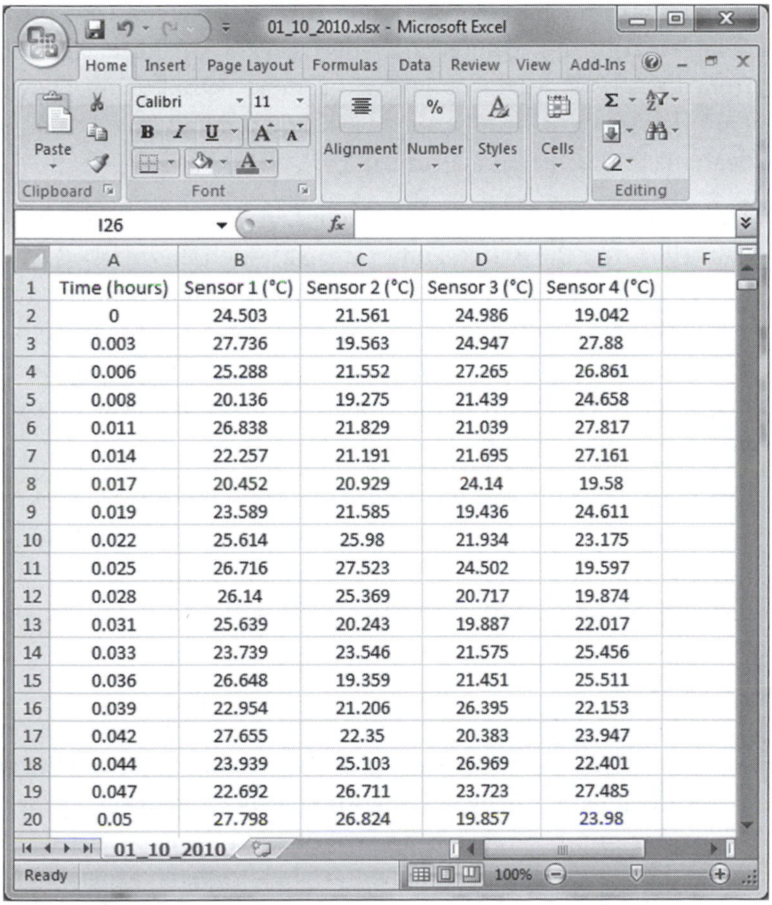

	A	B	C	D	E	F
1	Time (hours)	Sensor 1 (°C)	Sensor 2 (°C)	Sensor 3 (°C)	Sensor 4 (°C)	
2	0	24.503	21.561	24.986	19.042	
3	0.003	27.736	19.563	24.947	27.88	
4	0.006	25.288	21.552	27.265	26.861	
5	0.008	20.136	19.275	21.439	24.658	
6	0.011	26.838	21.829	21.039	27.817	
7	0.014	22.257	21.191	21.695	27.161	
8	0.017	20.452	20.929	24.14	19.58	
9	0.019	23.589	21.585	19.436	24.611	
10	0.022	25.614	25.98	21.934	23.175	
11	0.025	26.716	27.523	24.502	19.597	
12	0.028	26.14	25.369	20.717	19.874	
13	0.031	25.639	20.243	19.887	22.017	
14	0.033	23.739	23.546	21.575	25.456	
15	0.036	26.648	19.359	21.451	25.511	
16	0.039	22.954	21.206	26.395	22.153	
17	0.042	27.655	22.35	20.383	23.947	
18	0.044	23.939	25.103	26.969	22.401	
19	0.047	22.692	26.711	23.723	27.485	
20	0.05	27.798	26.824	19.857	23.98	

FIGURE 1.1 Example data file.

be charted over time and that some minimal trend analysis be performed within Excel using a trend line and best fit equation.

To perform the request manually in Excel for each file would be a very tedious task. Instead, it would be efficient to build an automation interface to Excel to complete the formatting dynamically. This would allow for batch processing of a group of files that need to be formatted in the same manner and would allow for flexibility of changing the report output if the reporting requirements were to change.

There are several methods of implementation from LabVIEW via the ActiveX interface. Three methods capable of achieving the same results will be discussed in this chapter. These solutions range in complexity and trade-off development effort with portability:

1. A modular and portable solution will be presented that provides complete control and automation of Excel and the Excel chart object from LabVIEW. This solution is the most complex but allows for LabVIEW to dynamically adapt to different data sets with no modifications to the virtual instrument (.vi).

2. Macros can be recorded and created in Excel and interfaced to LabVIEW over the ActiveX interface. Although this method requires less development effort, if changes to the reporting interface are necessary they will most likely need to occur to the macro and to the .vi. Macros also require certain security privileges to be enabled that may vary from system to system, which reduces portability of the program. This will be discussed in detail later.

3. Excel template (.xlt) files can be used to create the desired report format, and data can easily be linked to preconfigured data fields. This method also reduces development effort but requires changes to the template file and has other limitations as well.

SOLUTION

The first solution will discuss how to interface directly to Excel using ActiveX and the Excel chart object to plot the existing data from the given files programmatically. The workbook first needs to be opened programmatically and the number of rows and columns in the data set read. Then the data headers will be read and applied as labels to charts generated based on the size of the dataset. Then each column of data will be added as a new series on a separate chart and a trend line will be added to the plotted data.

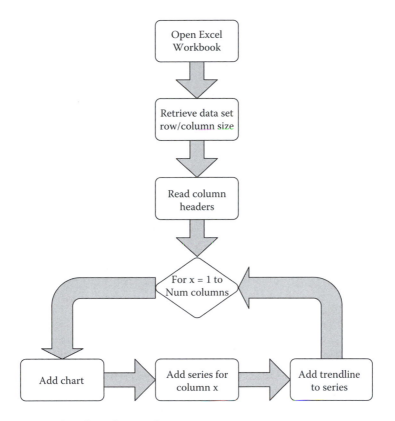

FIGURE 1.2   Flowchart for Excel ActiveX integration.

Note the data will not actually be read in by LabVIEW but only will be linked in the worksheet. Figure 1.2 presents a software flowchart.

Opening an Excel Workbook

To set things up, it is ideal to create a type definition cluster to store all the references that will be used in different portions of code or different SubVIs for ActiveX interaction. Ensure that references are managed and that none are left open inadvertently, which would cause memory leaks. Using a cluster to store these allows for easy passing of references to other parts of development code and provides for easy cleanup at the end of the ActiveX access. Another way to do this is to build the interface into a LabVIEW Class; however, this is out of the scope of this discussion.

References will need to be opened to the Excel application itself, the Excel workbooks, and a specific workbook (or file). References will be used later for sheets and charts within the file so these will be added to the cluster for

FIGURE 1.3   Structure of ActiveX references.

later use (Figure 1.3). Other references will also be used, such as for series; however, they will be managed completely inside the SubVI using them.

First, a reference to the Excel application itself will be opened. All ActiveX functions are located in the Connectivity>ActiveX palette (Figure 1.4). "Automation Open" returns a reference that points to the Excel ActiveX object (Figure 1.5).

To configure the Automation Open object for Excel, right-click on ActiveX Automation Open after it is placed on the block diagram and "Select ActiveX Class>Browse…." Then browse to the Microsoft Excel Object Library (Figure 1.6). Excel 2003 and 2007 correspond to Version 11

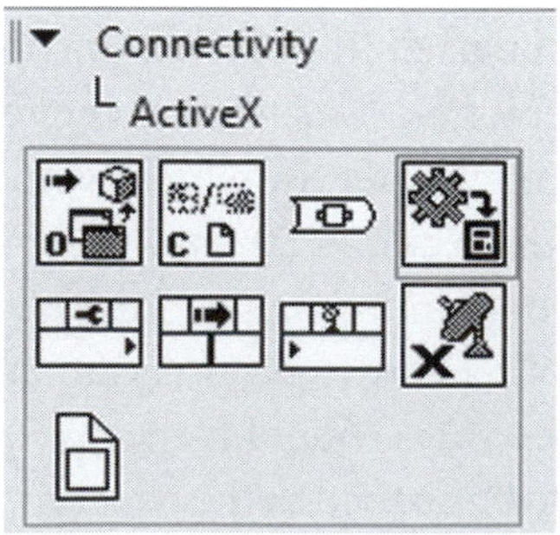

FIGURE 1.4   ActiveX palette.

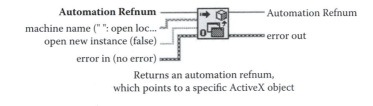

Returns an automation refnum,
which points to a specific ActiveX object

FIGURE 1.5   ActiveX automation open.

and Version 12, respectively. There are some differences in the ActiveX interfaces between the two versions, and they will be discussed later. Select the "Application" Object and press "Ok" to finish the configuration. Of course, Excel must be installed on the development and destination machines to use these solutions. ActiveX is merely providing a connection to the Excel application; it is not embedding the Excel source into the LabVIEW application.

To open a reference to a specific Excel workbook (or file), a property node on the "Application" reference is used to obtain a reference to "Workbooks." A property can also be added here to make Excel visible or hide it during automation. This is useful to keep applications looking professional. From the workbook reference use an "Open" method and specify the desired Excel file as input. The Open reference is synonymous with performing a "File>Open" in Excel, and it will open all files that are

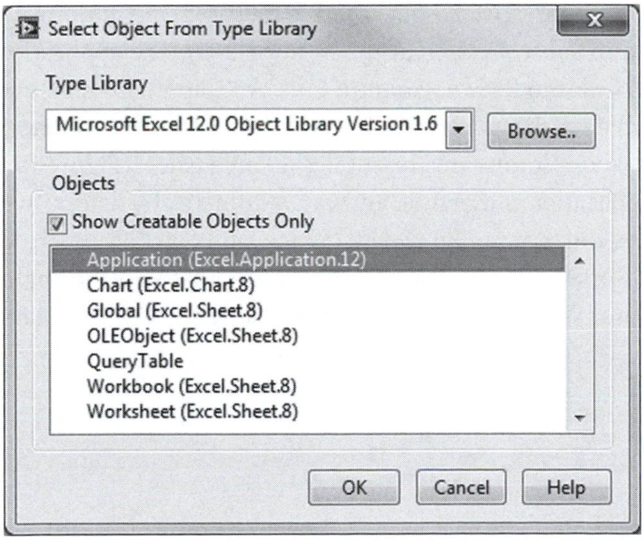

FIGURE 1.6   Select ActiveX class.

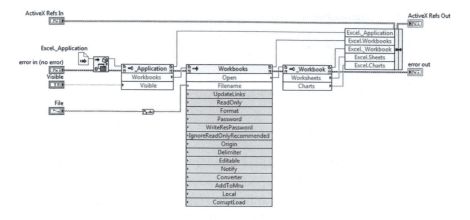

FIGURE 1.7 Open Excel and file.

readable by Excel (including tab-delimited files). A reference to the opened workbook is returned from the Open method, and this can be stored along with the Workbooks and Application references in the previously created cluster to pass on to later functions (Figure 1.7). References to the worksheets and charts in the opened workbook have also been saved.

Reading the Data Set Row and Column Size

Going back to the example, we would like to open the existing workbook and plot all sensor data on separate charts inside the workbook. As is shown in Figure 1.1, we are given a file with x-data (time) in Column A (or 1) and sensor data in subsequent columns B–E (or 2–5). The application requires charting of Column B vs. A, Column C vs. A, Column D vs. A, and so forth, each on separate charts. Before charting, it is necessary to determine how many rows and columns are present in the spreadsheet (Figure 1.8). In doing this, the application is dynamic and can automatically determine the number of charts and the amount of data in each based on how much data is present in the worksheet rather than hard-coding that into the application. To obtain the number of rows and columns in a worksheet, do the following.

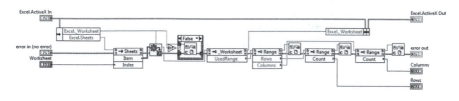

FIGURE 1.8 Retrieve worksheet size.

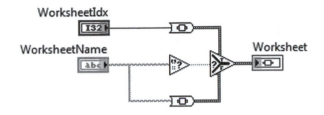

WorksheetIdx

WorksheetName

Worksheet

FIGURE 1.9   Index worksheet by name or index.

Specify the worksheet index for the worksheet that contains the dataset. Note that the worksheet indices start at 1 (not 0) for Excel and, furthermore, that chart tabs are not included in this index since we are indexing a worksheet here specifically. The worksheet could also be accessed by name instead of index by using a string as input to the Index terminal of the "Item" method. A dynamic way to accept either worksheet name or index is to accept both as input and then to check if the worksheet string input is blank. If it is, the worksheet index input is used instead of the worksheet name. This can be seen in Figure 1.9.

Notice that after indexing the worksheet the reference to it is closed only if one already exists; then the new reference is stored back onto the cluster. This makes the last accessed worksheet available for subsequent functions using the worksheet reference.

Reading the Column Headers

Next, read the column headers so we can title the charts and axes according to the datasets in the worksheet. This also helps to make the application dynamic such that the datasets can be renamed by the writing application without having to change this analysis application.

By specifying a worksheet index (as done previously to obtain the dataset size), the range of cells we would like to read back to retrieve the headers from the O.W. can be specified (Figure 1.10). This can be used later to specify the range of data for charting. The desired range is selected by specifying a start/end row and column (1 to n). The "Row Col to Range Format.vi" is used to convert the integer column indices to an ASCII alphanumeric string for Excel. Remember that Excel deals with rows and columns in the format [start column][start row]:[end column] [end row] (e.g., A1:E1000 specifies a two-dimensional range of columns A–E and rows 1–1000). First the correct worksheet is indexed and the cell range is obtained from the "Range" method. The "Cells" property then returns a reference to the cells defined by that range. An Invoke

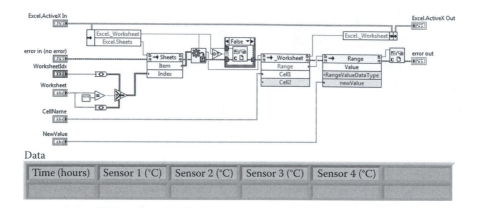

Data

Time (hours)	Sensor 1 (°C)	Sensor 2 (°C)	Sensor 3 (°C)	Sensor 4 (°C)	

FIGURE 1.10   Retrieve cell data.

node can then be used to return the data (as a variant) in those cells. Note: Many other operations on a range of cells can be performed at this invoke node such as Cut/Copy/Paste/Delete, Justify, Merge, Find, and Check Spelling. Variant data can then be cast to a 2-D array of strings to gain access to the header data. This will be used to add labels to the charts, axes, and worksheets. Notice again that after indexing the worksheet, the reference to that worksheet is closed only if one already exists, and the new reference is stored back onto the cluster for subsequent access to this worksheet.

Adding a Chart

Thus far we have opened our Excel file (workbook), selected the first worksheet and determined the bounding limits of the data, and have obtained the header information for the data. Now we want to add a new chart (on a separate worksheet) for each column of data. The first step is to actually add the chart tab or worksheet to our workbook using the "Add" method; then we will select the newly added chart, rename the sheet using the "Name" property, and finally invoke the chart wizard.

We can first obtain the count of the number of charts in the workbook. If this returns 0, then we simply add a chart (Figure 1.11a). If this returns more than 0, we can select the last chart in the workbook and add the chart after it (Figure 1.11b). Doing this will keep the newest chart right-most in the workbook tabs. We can also modify the chart type (e.g., bar, line, scattered) and then use the "Chart Wizard" to add labels to the new chart. The "Title" field of the chart wizard is the title that appears at the top of the chart, the "Category Title" is the x-axis label, the "Value Title" is

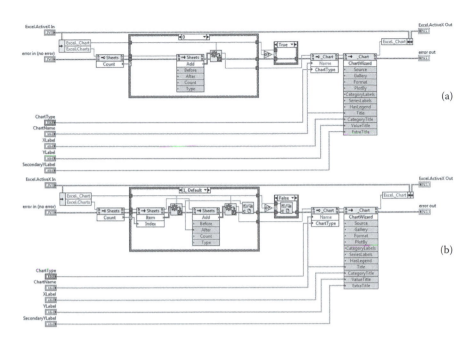

FIGURE 1.11   Add chart.

the y-axis label, and the "Extra Title" is the secondary axis label. All chart labels are optional and not required by Excel; however, naming the chart worksheet is required.

Note: In Excel 2007 when a new chart using the Chart Wizard is created, Excel automatically adds data from an existing worksheet if it can. Therefore, the automatically added data should be removed if using Excel 2007. The previously given code from Figure 1.11 can be modified to automatically handle this whether Excel has added the series' to the chart or not by simply obtaining the "Count" of series from the series collection after the chart is created and subsequently deleting those series (Figure 1.12).

Now the code in Figure 1.12 can be combined with the code that extracted the header information from the worksheet. The first column

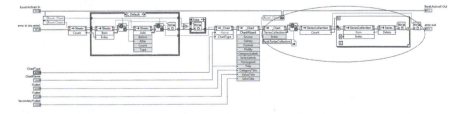

FIGURE 1.12   Add chart and delete auto-generated series.

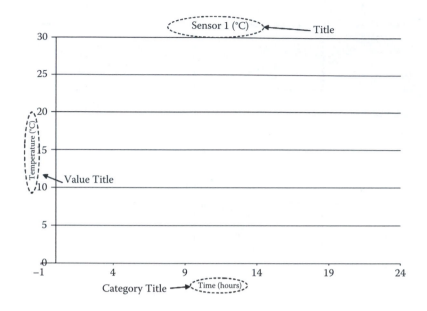

FIGURE 1.13   Excel chart labels.

header (Time) can be used for the x-axis label (Category Title), and the subsequent column headers (e.g., Sensor 1, Sensor 2) can be used for the chart name (Title). The y-axis label (Value Title) will be hard-coded for this example (Figure 1.13).

Add Series to Chart

Now that a chart has been created, a series must be added to the chart for every column of data to be added to the chart. For the example O.W., only one series for each chart will be added since we would like a separate chart for each column of data (or sensor). If all sensor data were plotted on the same chart, then the following process should be simply repeated for all columns of data on the same chart instead of on different charts.

First, open a reference to the worksheet where the data to be plotted will be drawn. In this example the worksheet was referenced by a reference index; however, it could also be referenced by name using a string as input to the Index terminal of the worksheet "Item" method as discussed previously. The worksheet reference will be used to obtain a reference to the range of x- and y-data to add to the series. Note that the O.W. is not actually being transferred from Excel to LabVIEW followed by transferring the data back to Excel using an OLE object, as with the Report Generation toolkit. Rather, we are simply telling Excel which

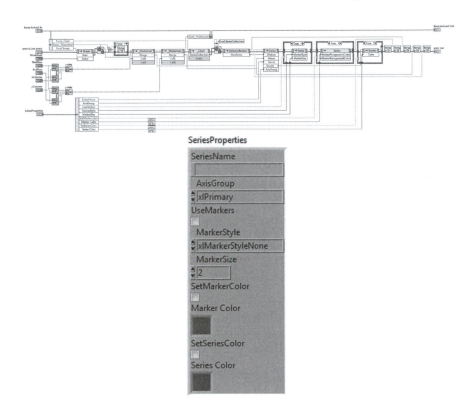

FIGURE 1.14   Add series.

cells to use for the series as if it were being manually selected in Excel using the Chart Wizard. This is much more efficient when dealing with large data sets and allows for direct linking to the existing data in the worksheet so that it can be easily modified from Excel later. Also, this method does not have a 4,000 data point restriction that is implied by OLE graph objects and will allow for plotting of up to 32,000 data points (Figure 1.14).

Once a reference has been obtained to the x- and y-data, a new series can be created and the references to the data and the name of the series (which will appear in the graph legend) can be passed in. Additionally, controls can be added to set the data to the primary or secondary axis (AxisGroup) and change the marker style, color, and the series line-color. Note that if the marker color properties are written, Excel will automatically turn on the markers, even if the marker style is set to "None." The "SeriesProperties" cluster in this example is a custom typedef containing some of the useful series properties that might be desirable to change. Any number of additional properties can be applied here including plot order, error bars, shadow, and even 3-D effects. Once the series settings have been finalized, it is important to close all references that have been previously opened. The reference to the series, the series collection, and the x- and y-ranges should all be closed.

Adding a Trend Line

Next, a trend line will be added to each series to help identify trends in the data. This could have been done at the previous step when the new series was added using the "NewSeries" reference. However, it will be explained how to index a previously created series instead. First, use the chart reference to gain access to the series collection for that chart. Remember that previously we stored the chart reference back to the Excel.ActiveX cluster; therefore, when accessing the chart reference here, it will point to the chart that was last accessed. Once a reference to the series collection has been obtained, we must index the correct series in the collection. For a example, there is only one series per chart so the "Series" index would just be 1. Remember that Excel indices start at 1 not 0. Once a reference to the series has been obtained, the "Trend lines" method can be used to get a reference to the trend lines for the series, and the "Add" method can be used to add a new one. The Type should be specified (e.g., exponential, linear, logarithmic), and it can be optionally named and display the equation on the chart. If "Name" is left blank, then Excel will automatically give it a default name of "Trend lineType(SeriesName)." It is important that the references to the trend line, the collection of trend lines, the series, and the collection of series are all closed as shown in Figure 1.15.

Note that both the series name and index can be accepted as inputs, which was discussed previously. If the series string input is blank, the series index input is used instead of the series name (Figure 1.16).

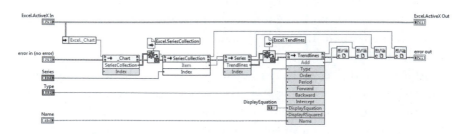

FIGURE 1.15   Add trend line.

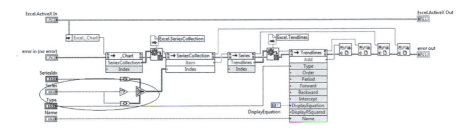

FIGURE 1.16   Add trend line series index or name.

Complete Integration for Example

The complete solution is shown in Figure 1.17. First open Excel to obtain a cluster of references and open the desired tab-delimited or Excel file. Then obtain the number of rows and columns in the data set on the first worksheet. Use the number of columns to read in all headers for each column of data. A for loop is then used that iterates through each column of data adding a chart with labels pulled from the previously read header data. A series of the column data and an appropriate trend line are also added to the chart. After all charts have been created, the for loop completes, and we close and save the Excel file, then close all open Excel references, and close the Excel application. The complete process would look something like Figure 1.17.

The resultant Excel workbook is shown in Figure 1.18.

Macros

Excel macros can also be used to automate Excel functions. This is achieved using Visual Basic for Applications (VBA). Excel has a macro recording tool that allows the recording of all steps as the user formats data as desired (Figure 1.19). After manually formatting, the user can stop the recording then view the recorded macro to see the VBA code that produces the same result as the manual control. The downside to this is that Excel's macro recorder will record every mouse click and event that is triggered by the user, and the resultant VBA code may be very inefficient. A further

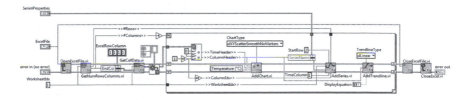

FIGURE 1.17   Complete example.

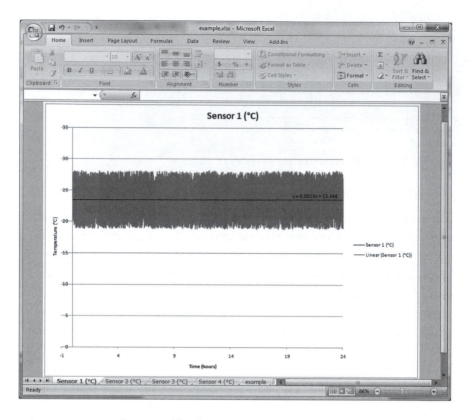

FIGURE 1.18   Resultant workbook.

consequence is that worksheets and cells are referenced by explicit name making portability of source code low. Furthermore, the recorder can even produce code that is nonfunctional in certain instances. So it is important to be able to understand the output VBA code and understand the effects of using the scripts as is and how to modify them for maximum portability.

The macro recorder can also be a useful tool to determine what methods and properties are necessary to interface to Excel in the desired way as in the previous section. It is quite useful to view a recorded macro to see what visual basic code reproduces the desired effect and then translate this into LabVIEW code. The interface is quite similar because LabVIEW and Visual Basic are using the same ActiveX interface.

Taking the input file and activating a recording, as is shown in Figure 1.19, will produce a VBA script something like what is shown in Figure 1.20 for adding a chart of the first sensor data only (column B versus column A in Figure 1.1). This example illustrates the need to understand the produced

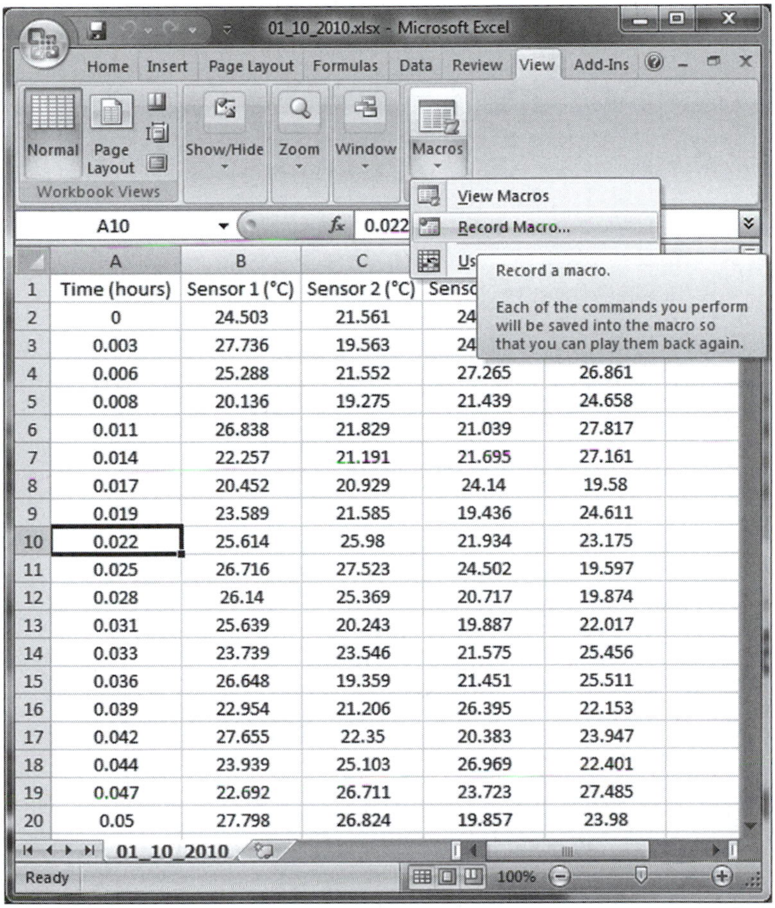

FIGURE 1.19   Excel's macro toolbar.

script before using it. As is shown on the first line, the Excel macro recorder references worksheets by worksheet name (which can change) instead of by their code names (which never change). So if the worksheet is renamed or if the macro is applied to another file that uses a different worksheet name, the macro will not be functional. This is also demonstrated by the chart name "Chart1" and the renaming of that chart to "Sensor 1(°C)" in Figure 1.20. Note that there are also numerous unnecessary mouse clicks moving back and forth from the data worksheet to the chart worksheet that recorded, which are identified by the ".Activate" properties. These are unnecessary and can be removed from the macro without affecting functionality.

```
Sub FormatChart1()
 Sheets("example").Select
 Charts.Add
 ActiveChart.SetSourceData
Source:=Range("'example'!$A$1:$E$8641")
 ActiveChart.ChartType = xlColumnClustered
 Charts.Add
 Sheets("Chart1").Select
 Sheets("Chart1").Name = "Sensor 1(°C)"
 ActiveSheet.ChartObjects("Chart 1").Activate
 ActiveChart.SeriesCollection(5).Select
 ActiveSheet.ChartObjects("Chart 1").Activate
 ActiveChart.PlotArea.Select
 ActiveSheet.ChartObjects("Chart 1").Activate
 ActiveSheet.ChartObjects("Chart 1").Activate
 ActiveChart.ChartType = xlXYScatterSmoothNoMarkers
 ActiveChart.PlotArea.Select
 ActiveChart.SeriesCollection(1).Delete
 ActiveChart.SeriesCollection(1).Delete
 ActiveChart.SeriesCollection(1).Delete
 ActiveChart.SeriesCollection(1).Delete
 ActiveChart.SeriesCollection(1).Delete
 ActiveChart.SeriesCollection.NewSeries
 ActiveChart.SeriesCollection(1).Name = "='example'!$B$1"
ActiveChart.SeriesCollection(1).XValues =
"='example'!$A$2:$A$8641"
 ActiveChart.SeriesCollection(1).Values =
"='example'!$B$2:$B$8641"
 ActiveChart.SeriesCollection(1).Select
 ActiveSheet.ChartObjects("Chart 1").Activate
 ActiveChart.SeriesCollection(1).Trendlines.Add
 ActiveSheet.ChartObjects("Chart 1").Activate
 ActiveChart.SeriesCollection(1).Trendlines(1).Select
 ActiveChart.SetElement (msoElementPrimaryValueAxisTitleRotated)
 ActiveSheet.ChartObjects("Chart 1").Activate
 ActiveChart.Paste
 ActiveChart.Axes(xlValue, xlPrimary).AxisTitle.Text =
"Temperature (°C)"
 ActiveSheet.ChartObjects("Chart 1").Activate
 ActiveChart.SetElement
(msoElementPrimaryCategoryAxisTitleAdjacentToAxis)
 ActiveSheet.ChartObjects("Chart 1").Activate
 ActiveChart.Axes(xlCategory, xlPrimary).AxisTitle.Text =
"Time (hours)"
End Sub
```

FIGURE 1.20   Formatting macro.

```
Sub Format1Chart()
 Charts.Add
 Sheets("Chart1").Select
 Sheets("Chart1").Name = "Sensor 1(°C)"
 ActiveChart.ChartType = xlXYScatterSmoothNoMarkers
 ActiveChart.SeriesCollection(1).Delete
 ActiveChart.SeriesCollection(1).Delete
 ActiveChart.SeriesCollection(1).Delete
 ActiveChart.SeriesCollection(1).Delete
 ActiveChart.SeriesCollection(1).Delete
 ActiveChart.SeriesCollection.NewSeries
 ActiveChart.SeriesCollection(1).Name =
"='example'!$B$1"
 ActiveChart.SeriesCollection(1).XValues =
"='example'!$A$2:$A$8641"
 ActiveChart.SeriesCollection(1).Values = "='example'!$B$2:$B$8641"
 ActiveChart.SeriesCollection(1).Trendlines.Add
 ActiveChart.SeriesCollection(1).Trendlines(1).Select
 Selection.DisplayEquation = True
 ActiveChart.SetElement
(msoElementPrimaryValueAxisTitleRotated)
 ActiveChart.Axes(xlValue, xlPrimary).AxisTitle.Text =
"Temperature (°C)"
 ActiveChart.SetElement
(msoElementPrimaryCategoryAxisTitleAdjacentToAxis)
 ActiveChart.Axes(xlCategory, xlPrimary).AxisTitle.Text =
"Time (hours)"
End Sub
```

FIGURE 1.21   Simplified formatting macro.

The previously recorded macro can be fine-tuned to remove unnecessary events. Also, due to issues with the macro recorder, the previously generated code will not actually function as intended without making necessary changes. All unnecessary event clicks to move between worksheets as well as the automatic series creation events were removed to produce the intended functioning macro as shown in Figure 1.21. Also note that the auto-generated series created by Excel when a chart is added to the worksheet must be deleted as was demonstrated in the first solution (Figure 1.12) in this chapter.

However, the previous macro still does not provide for a modular interface. Particularly, the x- and y-values for the graph are referenced by the worksheet name (which is derived off the file name automatically for our

```
Sub Format1Chart()
 Set xrange = ActiveSheet.Range("A2:A8641")
 Set yrange = ActiveSheet.Range("B2:B8641")
 Set xlabel = ActiveSheet.Range("A1")
 Set ylabel = ActiveSheet.Range("B1")
 Charts.Add
 Sheets(1).Name = xlabel
 ActiveChart.Name = ylabel
 ActiveChart.ChartType = xlXYScatterSmoothNoMarkers
 ActiveChart.ChartWizard Title:=ylabel, CategoryTitle:= "Time
  (hours)", ValueTitle:="Temperature (°C)"
 ' remove extra series
 Do Until ActiveChart.SeriesCollection.Count = 0
 ActiveChart.SeriesCollection(1).Delete
 Loop
 ActiveChart.SeriesCollection.NewSeries
 ActiveChart.SeriesCollection(1).Name = ylabel
 ActiveChart.SeriesCollection(1).Values = yrange
 ActiveChart.SeriesCollection(1).XValues = xrange
 ActiveChart.SeriesCollection(1).Trendlines.Add
 ActiveChart.SeriesCollection(1).Trendlines(1).Select
 Selection.DisplayEquation = True
End Sub
 Optimized Formatting Macro
```

FIGURE 1.22   Optimized formatting macro.

example). So, the previous macro would become nonfunctional if the data worksheet (or workbook) name were something other than "example." Furthermore, if this macro were to be applied to a workbook that has a different number of columns, the auto-generated series created by Excel would be different, and the macro would not delete the proper number of series. This may cause the macro to crash if it attempted to delete more series than what existed on the chart. This can easily be modified to make a more modular macro as is shown in Figure 1.22.

This macro sets the data range by calling out the active sheet instead of the worksheet by explicit name to reference the appropriate columns of data. Then the ChartWizard is used in much the same way that was demonstrated from LabVIEW directly in the first part of this solution. The ChartWizard can also be called from a macro to easily create a chart with the appropriate labels. This optimized macro will also automatically adapt to how many series exist on the newly created chart and will delete all existing ones until SeriesCollection.Count is 0 before adding a new series. The number of series automatically generated by Excel on a chart can vary; therefore, it is important that whatever method is used, it can adapt to any number of preexisting series present on the chart before adding additional series.

This macro could be further optimized to select the data range row limits based on the number of rows present in the worksheet, as was demonstrated in the previous LabVIEW solution. It can also be further modularized to handle all columns of data instead of just one as is shown here. This is left as an exercise to the reader.

Now that we have a functional and somewhat modular macro, how do we use it from LabVIEW with our example workbook? This becomes a two-part problem. First, the macro must be added to the workbook somehow. The initial problem statement was that the workbook is provided by another team, and it could be a tab-delimited file or an Excel workbook. It is assumed that it will not contain a macro to begin with (especially if it is a tab-delimited file, as these cannot store macros). Second, the macro must be run. Luckily there is a way to write a macro to a workbook programmatically from LabVIEW and even run it. How do we do it? You guessed it—through the ActiveX interface.

Beware that there are security permissions within Excel that need to be changed to allow adding and accessing macros from outside of Excel. Particularly in the Excel Trust Center the macro settings must be modified to trust access to the VBA project object model. This can cause issues with deploying code that uses macros to other workstations that do not have these permissions properly set within Excel. Of course, since this is a security setting, it cannot be modified programmatically. Therefore, macro implementation is not as ideal as the previous solution given that controls Excel directly from LabVIEW. Even so, the macro implementation will be continued for completeness.

First, the macro must be written using the same sequence of calls as was demonstrated earlier to open Excel and the specific workbook. Then, a reference to the workbook's VB Project must be opened and a new VB Component must be added. A simple way to do this is to add the code as a standard module in text format, as is shown in Figure 1.23; however, it can also be added from a file. Don't forget to close all the newly opened VB references after adding the module.

Once the macro has been added to the workbook, it can be run programmatically from LabVIEW using the "Run" method from the Excel

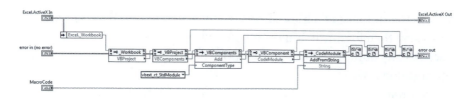

FIGURE 1.23   Adding macro to workbook.

application reference (Figure 1.24). Since the example macro does not have any arguments to be passed in, there is no need to wire any values to the argument inputs. If one were to implement a macro that takes arguments as input, then it would be necessary to wire those to the appropriate argument inputs. In this case, simply wire the macro name (Format1Chart) to the "Macro" input (Figure 1.25).

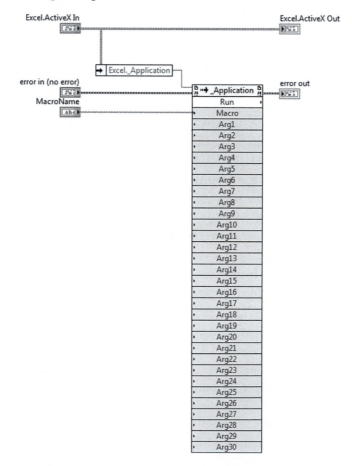

FIGURE 1.24   Running macro.

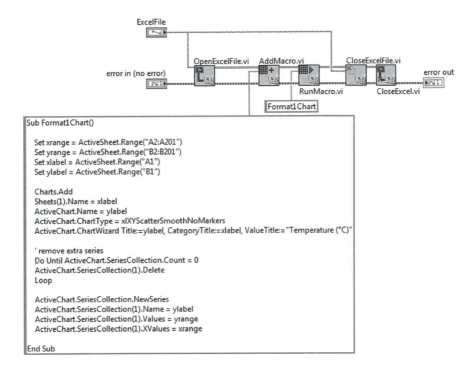

```
Sub Format1Chart()

    Set xrange = ActiveSheet.Range("A2:A201")
    Set yrange = ActiveSheet.Range("B2:B201")
    Set xlabel = ActiveSheet.Range("A1")
    Set ylabel = ActiveSheet.Range("B1")

    Charts.Add
    Sheets(1).Name = xlabel
    ActiveChart.Name = ylabel
    ActiveChart.ChartType = xlXYScatterSmoothNoMarkers
    ActiveChart.ChartWizard Title:=ylabel, CategoryTitle:=xlabel, ValueTitle:="Temperature (°C)"

    ' remove extra series
    Do Until ActiveChart.SeriesCollection.Count = 0
    ActiveChart.SeriesCollection(1).Delete
    Loop

    ActiveChart.SeriesCollection.NewSeries
    ActiveChart.SeriesCollection(1).Name = ylabel
    ActiveChart.SeriesCollection(1).Values = yrange
    ActiveChart.SeriesCollection(1).XValues = xrange

End Sub
```

FIGURE 1.25   Complete macro example.

Templates

Excel templates are spreadsheets that have already been preformatted for the desired look and feel and can also include required formulas and macros. They are saved as a special type (.xlt or .xltx) such that when one is opened a copy of the file is created in memory to preserve the contents of the template file. This can be quite useful in relation to our discussion since a complexly formatted spreadsheet where the graphs, formulas, and macros reference cells where the data will be located can be created. Then new data can be programmatically added to the preformatted spreadsheet by directly linking to custom cell names in the template. In this section the method to create a template for our example and add data into the template from LabVIEW will be shown.

An easy way to start a template is to start with a spreadsheet file that already has some sample data in it. For this example, the files as in Figure 1.1 will be used. The spreadsheet can then be formatted to look exactly how we would like the final output of each workbook to look. Charts and trend lines can be added for each sensor's data and the axes and chart labels can be formatted appropriately. Additionally, all formatting of cell widths,

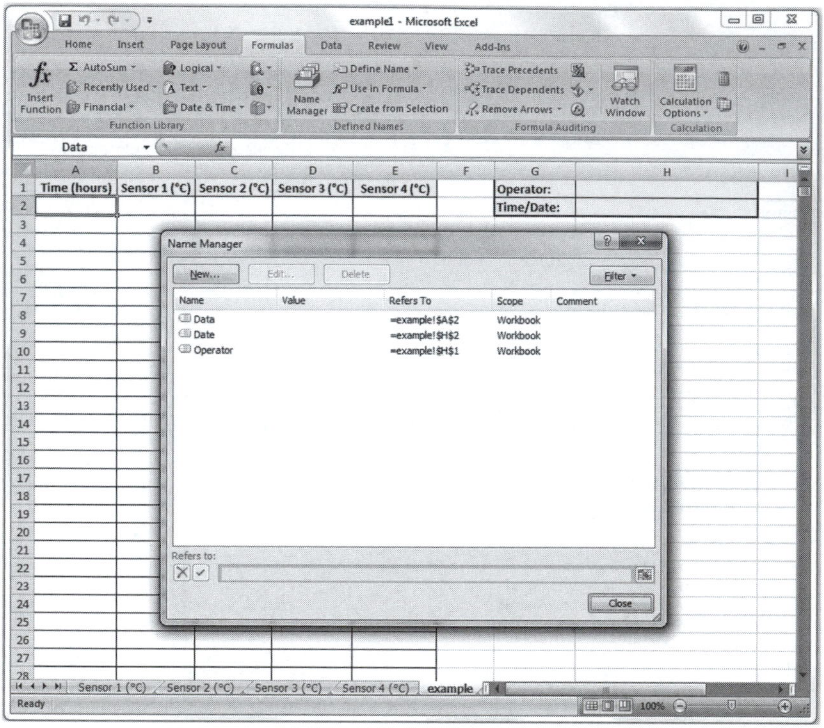

FIGURE 1.26   Setting up a template.

fonts, colors, and so forth will be preserved. To demonstrate this, some simple coloring and gridlines will be added to the data worksheet. Next, all the raw data such that only the column headers remain in the data worksheet can be deleted. The final step will be linking the cells such that one can easily insert data into the template programmatically. To do this, use the Name Manager (Formulas>Name Manager) in Excel to add names to the cells to be written to. The names "Data" for cell A2, "Operator" for cell H1, and "Date" for cell H2 as is shown in Figure 1.26 have been added.

Naming cells is optional as the data can be linked by referencing the cells explicitly by their reference (e.g., A2); however, adding a name to the cell location is a convenient way to improve portability since the location of desired data entry can be changed simply by changing the linked location of the cell name. The program inserting the data does not have to know where that location is; it simply references it by name, which doesn't have to change.

After formatting the workbook as it is to look, removing the raw data, and adding the appropriate cell names, the file needs to be saved as an Excel template by saving as an .xlt or .xltx file. Once it is saved, it can be

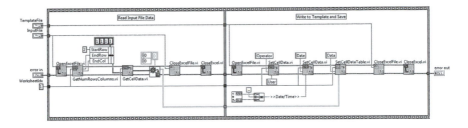

FIGURE 1.27   Template example using low-level ActiveX.

verified that opening the file will create a copy of the file in standard .xlsx format such that the original template file is preserved.

Once a template file is created, it can be opened programmatically and the desired data can be inserted using the named references created in the template. First, open the given tab-delimited or Excel workbook and extract the sensor data from the file. Then, insert the data into the template using ActiveX and the named cells. The cells where the data is stored can be read in the same way the headers were read in the first example (Figure 1.10), except that the data is read in as a float data type instead of a string. Then open the template previously created (which will automatically create a copy of the workbook in memory) and insert the 2-D float array of data. The complete example is shown in Figure 1.27. First the input file is opened, and the raw temperature data are extracted as a 2-D array of floating point values. Then the Operator, Date, and Data fields are populated with the relevant data using the functions shown in Figures 1.28 and 1.29.

To write the Data array (SetCellDataTable.vi in Figure 1.27), first obtain a reference to the "Data" named cell; then, using this as the origin of data entry, retrieve a subsequent range from this origin to the limits of the size of the input data set array. This will allow the "Value(put)" method to set the array starting from the index of the named cell "Data." This is shown in Figure 1.28.

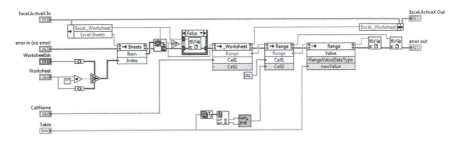

FIGURE 1.28   Write cell data as array.

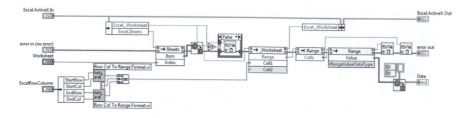

FIGURE 1.29   Write single-cell data as string.

We can do something similar for a single cell (SetCellData.vi in Figure 1.27) where only Cell1 is defined in the range input as the cell name and the value is set as a string. This will be used to set the operator and date fields. These fields were added here only for the purposes of demonstrating templates and linking to data cells.

If the Report Generation Toolkit is available, it also has a convenient Express VI that links to template files and automatically detects existing named cells. The MS Office Report Express VI is located in the Report Generation palette, and its configuration screen is shown in Figure 1.30.

FIGURE 1.30   MS Office Report Express VI configuration.

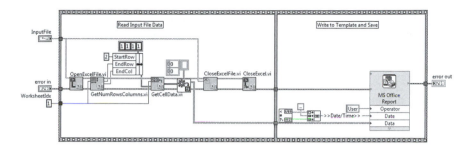

FIGURE 1.31 Template example using Report Generation Toolkit MS Office Report Express VI.

After selecting "Custom Report for Excel" in the Template dropdown, the path of the template file can be selected. Once selected, the Express VI automatically displays a list of detected named cells that data can be wired to. Figure 1.31 shows the implementation of the MS Office Express VI for the given template and example.

CONCLUSIONS

The first solution (Figure 1.17) showed how to fully automate Excel and some of its formatting features from LabVIEW using the ActiveX interface. This provided for a very flexible and scalable solution since all of Excel's features can be accessed remotely through LabVIEW without having to configure anything within the Excel application itself. This also proved to be the most time-consuming in terms of development time if all ActiveX functionality were developed from the ground up as was demonstrated.

As a second solution (Figures 1.19–1.25), the notion of macros was introduced. The use of the macro recorder within Excel proved to be a quick way to record desired functionality in a VBA script but also showed that it is prone to mistakes and must be fine-tuned for optimal portability.

Finally, the use of template files (Figures 1.26–1.31) was introduced to provide an even faster way to format Excel files in a particular way. This method also proved to be portable to a certain extent through the use of named cells. This method provides a nice balance between development time and portability.

Given all three solutions, the idea would be to create a top-level batch processing engine to format hundreds of these input files in the exact same way and also to provide some level of portability if formatting requirements were to change in the future (e.g., more sensors added, additional data graphed simultaneously). It is up to the developer to weigh the advantages and disadvantages of each method to determine which one is an optimal solution for the implementation. For a more complete library of Excel ActiveX automation functions please visit http://www.aledyne.com/products.html.

RESOURCES

LabVIEW Development System 8.6 or Higher (for Windows)

LabVIEW Report Generation Toolkit

Microsoft Excel 2003 and 2007.

Interacting with Windows Applications

Bill VanArsdale

CONTENTS

INTRODUCTION

An important task for many LabVIEW programs is interacting with other applications on the same computer. This interaction is sometimes necessary to accomplish tasks not readily handled within LabVIEW. The method of interaction can be indirect, such as changing saved settings. However, most tasks require more direct control. Such control is available in Windows through a command line interface as well as properties and methods associated with ActiveX and .NET objects. This chapter will focus on interacting with other Windows applications using these approaches within the LabVIEW environment.

This chapter employs specific notation and provides many diagram images to illustrate methods for interacting with Windows applications. While figures are not intended to document example VIs, I find such Images the most helpful feature in a LabVIEW presentation. I also try to employ the following notational convention throughout this chapter:

italic: paths, files, example VIs and subVIs, URL

"quotes": LabVIEW variable names in VIs and subVIs

bold: commands, menus, LabVIEW development system VIs

`courier`: command-line inputs and VBScript

Hopefully the chapter figures and notations are useful to your understanding of this material. Example VIs and other materials are available through the book's web site.

BACKGROUND

All versions of the LabVIEW development system provide useful connectivity VIs and functions as shown in Figure 2.1. **ActiveX** functions get and set properties and invoke methods with ActiveX objects such as Microsoft Office applications. Similarly, **.NET** functions access properties and methods for .NET objects such as those found in the Microsoft .NET Framework. The **Libraries & Executables** subpalette contains functions for calling DLLs and shared libraries, interfacing with text-based languages, and executing a system command. The **Windows Registry Access VIs** are used here to change application settings. The first seven solutions involve VIs that can be opened and edited in any LabVIEW development system, version 8.5 or later. VIs for Solutions 8 and 9 should be editable on all but the Base version of LabVIEW 8.5. This version does not include

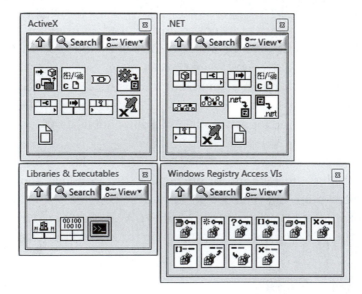

FIGURE 2.1   Useful subpalettes available on the Connectivity VIs and Functions palette.

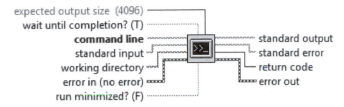

FIGURE 2.2 The *System Exec.vi* allows LabVIEW applications to send commands and receive output from the Windows command interpreter.

the **Event Structure** as well as functions **Register Event Callback** and **Unregister For Events** on the **ActiveX** and **.NET** palettes.

Several VIs described in subsequent sections use the **System Exec.vi** with the connector pane shown in Figure 2.2. This VI enables LabVIEW applications to communicate with the Windows command interpreter. Launch this program by typing "cmd" in the **Run...** dialog available through the Windows **Start** menu or open the executable *cmd.exe* typically found in the folder *C:\Windows\System32*. A variety of tasks can be handled directly through this interpreter using commands supported by the Windows operating system. A list of commands available for Windows XP can be obtained by opening **Help and Support** through the Windows **Start** menu and searching for "a-z." A good stand-alone reference is the help file *WinCmdRef.chm*, which can be downloaded from Microsoft. Other references are available online by typing "windows command line" into an available search engine.

Additional connectivity is provided by **Report Generation** and **Database** VIs included with LabVIEW add-on toolkits. These toolkits, available separately or as part of the NI Developer Suite Core Package, add the subpalettes shown in Figure 2.3. While basic Hyper Text Markup Language (HTML) and standard report VIs are provided in all LabVIEW development systems, the ActiveX-based tools for working specifically with Microsoft Excel and Word are available only as part of the Report Generation Toolkit. The Database Connectivity Toolkit VIs facilitate data exchange with common database applications such as Microsoft Access and SQL Server. VIs for the last three solutions will not run without installing the appropriate toolkit.

The following solutions are implemented as LabVIEW 8.5 VIs. It is important to note that some of these VIs require structures, functions, and toolkit VIs not included with all LabVIEW development systems. VIs designed to

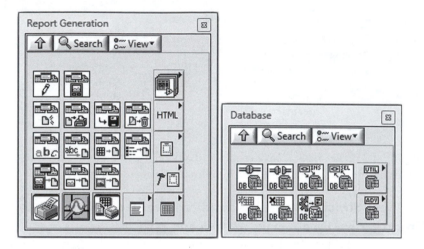

FIGURE 2.3   Report Generation VIs in the Report Generation Toolkit and database VIs in the Database Connectivity Toolkit.

interact with freely downloaded executables or commercial applications like WinZip, PowerPoint, Excel, and Word will not work without that software installed on your computer. These VIs have been tested using Office 2003 and 2007 applications and WinZip 9 and WinZip Pro 12. All VIs should be able to be opened and saved in more recent versions of LabVIEW.

All solution VIs have been developed and tested on Dell computers with Intel processors using Windows XP Professional and Vista Ultimate. Other versions of these operating systems may not contain all commands used by the included VIs. If an error is encountered using a Windows command, try searching the web for this command by name. A missing command can often be downloaded from Microsoft or another site to the *C:\Windows\System32* folder, which will enable the desired functionality. Additional instructions are provided in Solution 3 and the Resources section regarding the Microsoft .NET Framework.

Some of the solution VIs have been analyzed using **VI Analyzer**. This LabVIEW utility, available as part of the VI Analyzer Toolkit, provides feedback on wiring style and error testing. Since all solution diagrams are somewhat consistent with the humorous dictum, "White space is evil, destroy nearly all of it" (Conway and Watts, 2003, p. 132), the financial trader's wall of monitors will not be needed to view them in their entirety. However, **VI Analyzer** basically dislikes the occasional wire under objects and less than square icons with error terminals not in the absolute lower left and lower right positions. It is the author's hope that readers will separate content from implicit style suggestions in the following presentation.

CHALLENGE 1

Controlling the Windows Screen Saver

Screen saver activation can hamper operations as well as often being inconvenient. Background tasks like the screen saver can also adversely influence real-time applications and data acquisition routines and can slow down applications that require immediate responses to specified conditions. The challenge here is to control how and when a screen saver launches. These tasks can be performed programmatically using *set screen saver.vi* and *start screen saver.vi* described in the following solution.

SOLUTION 1

Controlling the Windows Screen Saver

The Windows Screen Saver stores settings for each user in the Windows Registry. These settings are normally specified using the **Screen Saver** tab on the **Display Properties** dialog. This control panel may be accessed by right-clicking on the desktop and selecting either **Properties** (XP) or **Personalize** (Vista and Windows 7). These settings can be changed programmatically using *set screen saver.vi*, shown in Figure 2.4. This VI writes changes to the registry only by comparing to the current entry. Use the **Registry Editor** to view changes to screen saver settings. Launch this program by typing "regedit" in the **Run...** dialog available through the Windows **Start** menu, or open *C:\Windows\System32\regedt32.exe*. Exercise unusual caution when modifying registry entries. Remember: You break it, you buy it.

Screen saver files with extension *.scr* are located in the *C:\Windows\ System32* directory. Double-clicking any of these files starts a screen saver. Starting these files using the **System Exec.vi** requires the "command line" input filename.scr /d as shown in the *start screen saver.vi* diagram in Figure 2.5. This VI waits for the screen saver to deactivate only if "wait?" is true.

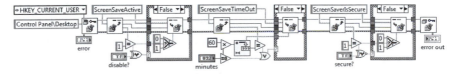

FIGURE 2.4   The *set screen saver.vi.*

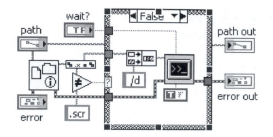

FIGURE 2.5   The *start screen saver.vi.*

This VI and several others use the custom subVI *get directory, file name and extension.vi* to extract the file name and extension from "path." An error is produced if "path" is either a directory, empty, invalid, or refers to a file with an extension other than *.scr*. This type of error checking is typical of command line VIs described in subsequent solutions. Note that **VI Analyzer** would generate a diagram style error for not enclosing all subVIs in a case structure to pass an input error. Such an approach would slightly speed execution of the Error case at the expense of the No Error case.

CHALLENGE 2

Opening Files and Folders

Opening files and folders from a LabVIEW application is often important. For example, any *.ini* file could be opened in WordPad to allow users to edit and save keys. A data folder can be opened to display currently available files in Windows Explorer. The Windows Paint program can be opened to allow editing of an acquired image. The challenge here is to open both files and folders, where files are displayed in an associated or specified application. These tasks can be performed programmatically using *open file or folder.vi* described in Solution 2.

SOLUTION 2

Opening Files and Folders

The "command line" input cmd.exe /c app "path" to **System Exec.vi** is used to open a file with a known application. Such applications reside in directories specified by the PATH environment variable. Typing a path at a command prompt provides a list of these directories. For example, applications such as *calc.exe* (Calculator), *explorer.exe* (Windows Explorer), *mspaint.exe* (Paint), *notepad.exe* (Notepad), *taskmgr.exe* (Windows Task Manager), and *write.exe*

(WordPad) typically reside in the *C:\Windows\System32* directory. Consequently, the command cmd.exe /c notepad.exe "C:\ Users\Public\ReadMe.txt" would open the *ReadMe.txt* file in Notepad. A similar command would open an image file in Paint or a folder with Windows Explorer.

If app is unspecified, files open in the application specified by the file type associated with the extension. For example, a file with the extension *.txt* would typically open in Notepad. A Windows dialog requesting the user to supply an application will appear if the extension is an unrecognized file type. Files without an extension should be opened using *explorer.exe* if no other program is specified. In this case, the Windows **Open With** dialog appears, providing the user with a list of installed programs. Such dialogs should be invoked only by interactive applications.

While Microsoft Office files are easily displayed using this method, these applications do not allow other file types to be opened. For example, a text file could not be opened using Word or Excel. A trick for opening a tab-delimited text file in Excel is to change the extension to *.xls* and then open the file. A similar method can be used with Word by changing the text file extension to *.doc*. ActiveX properties and methods are later used to accomplish other tasks with these applications.

Finally, this command opens a specified application if "path" is excluded. Other applications can be opened through their executable path or a shortcut that points to that file. For example, Office applications can be opened through the shortcut file in a directory associated with the Windows **Start** menu. Right-click on this menu and select **Explore All Users** to find this directory. If both app and "path" are unspecified, the Windows command opens the *Computer* (root) directory.

The *open file or folder.vi* diagram shown in Figure 2.6 provides a method for opening a file at "path" in a specified "app." The other case

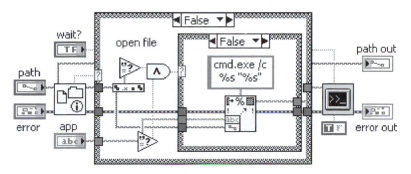

FIGURE 2.6   The *open file or folder.vi*.

opens a folder or known application. This VI waits for the application to close if "wait?" is true.

CHALLENGE 3

Checking File Properties

LabVIEW provides the **File/Directory Info** function on the **File I/O>Advanced File Functions** subpalette to determine the last modification date and time as well as the size of a file or number of files in a directory. This function is used in the previous example to determine if "path" is a directory. However, files have many properties that are not detected by the **File/Directory Info** function. For example, right-clicking on a file and selecting **Properties** will display information on **Read-only** and **Hidden** attributes on the **General** tab as well as version information for most executables on the **Details** tab. The challenge here is to read and set file attributes as well as to obtain version information for executable files.

SOLUTION 3

Checking File Properties

The attrib command reads and sets attributes for a specified file using the method shown in Figure 2.7. This information is output as the string "standard output" from **System Exec.vi**. The default value of the input, "wait until completion?(T)," must be used to obtain output from this VI. The output contains capital letters such as R: read-only and H: hidden, denoting set attributes. In addition, switches ±r (read-only) and ±h (hidden) can be used to set (+) or

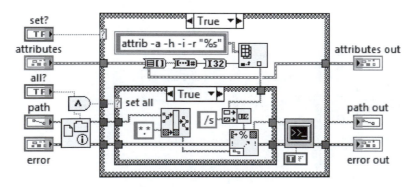

FIGURE 2.7   The *file attributes.vi*.

clear (−) attributes for a specified file. Again, the key here is to "wait for it" with regard to "standard output" from **System Exec.vi**.

The *file attributes.vi* diagram shows a method for setting archive, hidden, index, and read-only "attributes" for a specified file at "path." These "attributes" are applied to all files in a specified folder, and its subdirectories if "all?" is true. Attributes for a file at "path" are output as "attributes out" if "set?" is false.

The diagram shown in Figure 2.8 uses a .NET method to obtain version information for all executables in a folder. To generate code that produces similar results, place an **Invoke Node** on a diagram and select the right-click menu item **Select Class>.NET>Browse...** to obtain the **Select Object From Assembly** dialog. Select the .NET Assembly **System(2.0.0.0)** and Object **System.Diagnostic. FileVersionInfo**. Left-click on the lower part of the **Invoke Node** and select the method **[S]GetVersionInfo(String fileName)**. Download and install the most recent version of the Microsoft .NET Framework if this method does not appear on your system. Additional .NET examples, including SimpleTaskMonitor.vi, are available through the **NI Example Finder** (select **Find Examples...** from the **Help** menu in LabVIEW and navigate to Communicating with External Applications>.NET).

The *get file version info.vi* diagram (Figure 2.8) obtains version "info" for all files with specified "extension" in "folder." Alternatively, setting "extension" to "*" provides "info" for all files in "folder." While nodes in Figure 2.8 are displayed in the **Short Names** format, the **No Names** format is used for properties and methods in subsequent solutions. Right-clicking on any node and selecting **Name Format>** will change the appearance. Hover with the wiring tool to obtain terminal names in tip strips for nodes in any format.

Another method of obtaining version information uses the LabVIEW 2009 VI library *fileVersionInfo.llb* typically located in *C:\ Program Files\National Instruments\LabVIEW 2009\vi.lib\Platform*. These VIs call *kernel32.dll* and *version.dll* using the **Call Library**

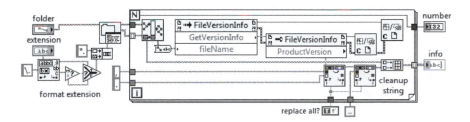
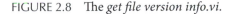

FIGURE 2.8 The *get file version info.vi*.

Function Node available on the **Libraries & Executables** subpalette shown in Figure 2.1. These application files usually reside in the *C:\Windows\System32* directory. In LabVIEW 2009, select **Save for Previous Version...** from the **File** menu of *FileVersionInfo.vi* to save this library to LabVIEW 8.5.

CHALLENGE 4

Creating Shortcuts

Windows provides a method for pointing to other files through a shortcut or link (*.lnk*) file. The binary contents of a link file can be viewed by dropping it onto an open Notepad window. The path to the executable and other information should be discernable. While shortcuts are easily created by right-clicking on most Windows files, no standard command is available in Windows XP or Vista for creating a link file from the command prompt. The challenge here is to programmatically create shortcuts, including the specification of run style window and comment.

SOLUTION 4

Creating Shortcuts

If a command is not available in Windows, someone has usually created an executable with a command-line interface to address the deficiency. For example, *shortcut.exe* (Optimumx) creates link files from the command prompt. While such applications can be called by name if placed in the *C:\Windows\System32* directory, the diagram in Figure 2.9 calls this executable by path. This VI is a simple example of adding functionality through executables that are readily available online. Another example is *truecrypt.exe* (TrueCrypt), which

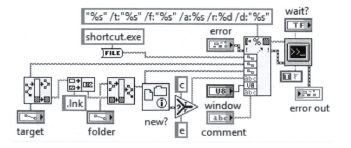

FIGURE 2.9 The *create shortcut.vi*.

provides on-the-fly encryption through a command line interface. Many of these executables, including *shortcut.exe* and *truecrypt.exe*, are available free of charge (freeware).

The *create shortcut.vi* diagram is a method for calling *shortcut.exe* residing in the same directory. This application uses switches (t, f, a, r, d) defined in the *ReadMe.txt* file included with the download. A *.lnk* file with the "target" file name is created in "folder" with specified run style "window" and "comment." The subVI *path to file in directory.vi* is used to create a path to *shortcut.exe* in the same folder as the subVI. This VI will not successfully run without downloading *shortcut.exe* to that directory.

Another method of creating a shortcut is to call the VBScript *shortcut.vbs* using the diagram shown in Figure 2.10. This script file is obtained by copying the following text:

```
Set WshShell = WScript.CreateObject("WScript.Shell")
Set lnk = WshShell.CreateShortcut(Wscript.Arguments.
Named("F"))
lnk.TargetPath = Wscript.Arguments.Named("T")
lnk.Description = Wscript.Arguments.Named("D")
lnk.WindowStyle = Wscript.Arguments.Named("R")
lnk.Save
```

into Notepad and saving as *shortcut.vbs* (set **Save as type:** to **All Files**). This script uses the object CreateShortcut (*http://ss64.com/vb/shortcut.html*), where case-insensitive command switches (F, T, D, R) are defined to mirror *shortcut.exe*. This script is called by path using the cscript command in Windows. Other VBScript examples can be found online and in the *C:\Windows\System32* directory. (Note: Do not modify these scripts.)

The *create shortcut using script.vi* diagram shown in Figure 2.10 is a method for calling *shortcut.vbs* residing in the same directory. This VI, like the previous example, uses the custom subVI *path to file in directory.vi* to get a path to the specified file "name" in the same

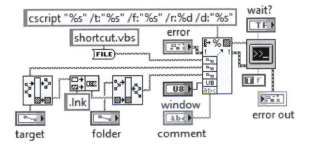

FIGURE 2.10 The *create shortcut using script.vi.*

directory. A *.lnk* file with the "target" file name is created in "folder" with specified run style "window" and "comment." This VI will not run successfully without copying *shortcut.vbs* to the same directory as the subVI *path to file in directory.vi*.

CHALLENGE 5

Working with WinZip

Zip archives are containers for files similar to folders and VI libraries. Just as Windows Explorer and the LabVIEW LLB Manager display and manipulate the contents of folders and VI libraries, the WinZip application is commercial software that provides similar capabilities for zip archives. While LabVIEW 8.5 includes VIs for adding to a zip file and unzipping an archive, a richer set of tools can be obtained through WinZip. Files added to an archive can be compressed and encrypted using various algorithms. File attributes can be retained or removed during this process. Text comments for an individual file or all files can be edited and stored with the archive. Files can be displayed using internal viewers or external applications. Files can be edited within an archive and then extracted to a specified directory. The challenge here is to use WinZip's functionality to search files within an archive to determine which contain a specified string.

SOLUTION 5

Working with WinZip

Use the command `findstr /s /i /m /c:"string" *.*` with **System Exec.vi** to search for all files (*.**) in a specified "working directory" containing the literal (`/c:`) "string". This case-insensitive (`/i`) search checks all subfolders (`/s`) to return a list of file names (`/m`) as shown in Figure 2.11. This VI can find words in *.txt*, *.rtf*, and *.doc* files but not *.pdf* and other purely binary files. The search also finds instances within zip archives, but a subVI using WinZip is needed to determine which archived files contain "string."

The *search folder for string.vi* diagram in Figure 2.11 is a method for searching all files of specified "type" within "folder" for "string." Note that "folder" is connected to the "working directory" terminal of **System Exec.vi**. A subVI *search zip archive for string.vi* is used to search within a zip archive.

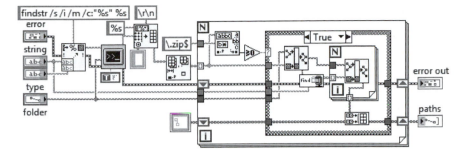

FIGURE 2.11   The *search folder for string.vi*.

LabVIEW interacts with WinZip through a command line interface available as the WinZip Command Line Support Add-On. This package provides executables *WZZIP.exe* and *WZUNZIP.exe* for zipping and unzipping files. These executables are part of a commercial package that can be installed for an evaluation period on your computer. The *WZUNZIP.exe* application determines a list of files in an archive and can individually extract each file to the console. These features allow users to search each file's contents for a specified "string" as shown in Figure 2.12. While *.zip* files contained within "archive" are also searched, the VI cannot determine which elements of these files contain "string." This level of recursion would require extracting *.zip* files within "archive."

Figure 2.12 shows the *search zip archive for string.vi*, which is a method for searching within a specified "archive." Much of this diagram involves output string processing to obtain a file list. These files are individually extracted to the console and then searched for the desired "string." This VI requires *WZUNZIP.exe* to reside in the same directory as the subVI *path to file in directory.vi* or in the same folder as an executable containing this subVI.

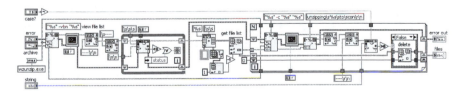

FIGURE 2.12   The *search zip archive for string.vi*.

CHALLENGE 6

Logging Application Events

Error and warning dialogs are usually not appropriate for continuously running applications. Such dialogs stop at least one program loop until closed by the user. These applications require that errors and warnings are logged with minimal interruption. While text files are a reasonable repository for this information, Windows applications usually write error, warning, and information events to an Application Log shown in the **Event Viewer**. This program, typically located at *C:\Windows\System32\eventvwr.exe*, is often available under *Administrative Tools* in the Windows XP **Start** menu. System administrators use the **Event Viewer** to monitor and process Windows Event Logs. The challenge here is to log error, warning, and information events programmatically to the Application Event Log.

SOLUTION 6

Logging Application Events

The Windows command `eventcreate /l APPLICATION / so "source" /t {ERROR,WARNING,INFORMATION} /id eventID /d "description"` adds an error, warning, or information event to the Application Log. This command is used in the *log application event with cmd.vi* diagram shown in Figure 2.13. The `eventID` has the range 1–65535 typically associated with U16 integers. Unfortunately, this command does not seem to log events unless `eventID` is in the range of 1–1000. Another version of this VI is also shown in Figure 2.13 but instead uses the .NET object **System.Diagnostics.EventLog** to accomplish the same task

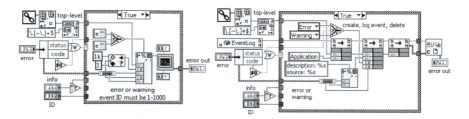

FIGURE 2.13 The log application event with *cmd.vi* and log application event with *net.vi*.

without this limitation. Errors and warnings are obtained from "error", while information events specified by "info" and "ID" are handled in the hidden False case.

The subVI *error description.vi* is used in this solution to extract the code, description, and source from the input "error." This subVI uses **Error Code Database.vi** typically located in *C:\Program Files\ National Instruments\LabVIEW 8.5\vi.lib\Utility\error.llb* to obtain error descriptions for most codes. Descriptions for custom error codes in the range 5000 to 9999 and –8999 to –8000 are provided by a cluster control containing code and description arrays. VI properties and methods can be used to programmatically update such a control in Edit Mode and then set the default value. LabVIEW offers other methods for defining custom error messages.

Another VI included with this solution opens the Windows **Event Viewer**. Run this VI or launch the executable to view the Application Event Log. Use both VIs to create test events and then refresh the Application Log. Double-click or right-click on an entry to open the **Event Properties** window. The Description field will contain a disclaimer stating that the source is unregistered if generated using the .NET approach. This process works well under Windows XP with somewhat normal permissions. However, one must right-click on a LabVIEW shortcut and select **Run as administrator** to successfully log events with either solution VI under Vista.

CHALLENGE 7

Monitoring System Resources

LabVIEW provides the name and version of the operating system running on the local computer through the **Property Node** on the **Application Control** palette. The default class is **Application**, but right-clicking near the top of any **Property Node** and selecting **Select Class>VI Server>Application** from the menu will obtain this setting. Left-click on "Property" or the lower part of the node to obtain **Operating System>Name** or **Operating System>Version Number**. Some applications may require additional system information such as memory size. This information is displayed on the **General** tab of the **Properties** dialog obtained by right-clicking on **My Computer** either on the desktop or under the Windows **Start** menu. Information on other computers connected to your system may also be desired. The challenge here is to obtain system information programmatically for the local machine and other networked computers.

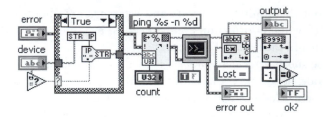

FIGURE 2.14 The *ping network device.vi.*

SOLUTION 7

Monitoring System Resources

Windows provides useful commands for monitoring system resources. Some of these commands are listed as follows:

ping <"device"> -n <"count"> (sends "count" echo request messages to "device" and checks for replies)
ipconFigure 2-/all (displays the full TCP/IP configuration for all adapters)
netstat -n (displays all active TCP connections with addresses in dot notation)
systeminfo /s <"computer"> /fo csv (displays configuration information for "computer" in CSV format)
tasklist /s <"computer"> (displays a list of currently running processes on "computer")

Methods for using these commands from LabVIEW are shown in diagrams for *ping network device.vi* in Figure 2.14 and *get sysinfo.vi* in Figure 2.15.

The *ping network device.vi* diagram in Figure 2.14 shows a method for sending "count" echo request messages to "device." The local computer is pinged if "device" is empty. Such pings are usually successful unless your computer does not have a network adapter. The indicator "ok?" is true if every message generates a reply.

Figure 2.15 presents a method for obtaining a list of executables currently active on "computer." Information for the local machine is

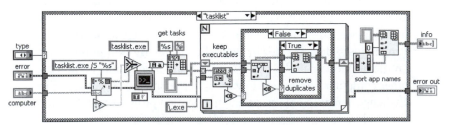

FIGURE 2.15 The *get sysinfo.vi.*

obtained if "computer" is unspecified. This information is output as the first column of "info" for the `tasklist` command specified by "type."

This VI implements the last four commands using the "type" enum for specification. Processing involves waiting for the "standard output" from **System Exec.vi** and then parsing a string intended for the Windows command interpreter. Parsing typically involves the **Match Pattern**, **Search and Replace String**, and **Spreadsheet String to Array** functions on the **String** palette to extract useful information as an array.

The Internet Toolkit for LabVIEW provides additional VIs for communicating with networked devices. In particular, this toolkit contains Telnet VIs for logging into another network device from a local computer. This network protocol is adequate for an isolated or private network (intranet) where security is less of an issue. However, many system administrators require a more secure protocol, such as the Secure Shell (SSH), for "snooping around" their network. SSH can be implemented on LabVIEW using a command line version of PuTTY. The PuTTY Link application *Plink.exe* is useful for securely initiating automated operations from within LabVIEW.

CHALLENGE 8

Using Voice Commands

I have been interested in speech recognition since watching Harrison Ford navigate an image using voice commands in the 1982 movie *Blade Runner*. Speech recognition and synthesis capability for Windows have been under development at Microsoft since the early 1990s. This technology allows one to control Windows and other programs using voice commands spoken through a microphone connected to a computer. The challenge here is to process voice commands using objects from the **System.Speech.Recognition** class in Microsoft's .NET Framework.

SOLUTION 8

Using Voice Commands

Windows Vista provides integrated support for speech recognition, whereas XP requires installation of the Speech Application Programming Interface (SAPI). An XP computer already has this capability if a **Speech Recognition** tab appears in the **Speech Properties** dialog obtained by selecting **Speech** from the **Control Panel** on the Windows **Start** menu. Download and install SAPI 5.1 on your Windows XP computer if speech recognition is not currently

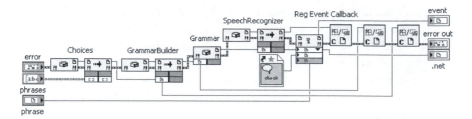

FIGURE 2.16   The *speech setup.vi.*

available as well as the current Microsoft .NET Framework if version 3.0 or higher is unavailable on your computer.

Select **Speech Recognition Options** (Vista) or **Speech** (XP) from the **Control Panel** on the Windows **Start** menu to open the **Speech Properties** window. Specify user settings, microphone configuration, and speech profiles on the **Speech Recognition** tab. The best type of microphone for this purpose is typically part of a headset. The user should train a speech profile with this device.

A method for establishing a set of voice commands is shown in Figure 2.16. This approach uses the **Register Event Callback** function to couple recognition with a value change in the control associated with the "phrase" reference. Consequently, a Value Change event for this control will register in an **Event Structure** upon recognition of a voice command.

The *speech setup.vi* diagram uses .NET objects from the **System. Speech.Recognition** class to build a grammar based on "phrases." This approach was demonstrated by Leo Cordaro (*http://decibel. ni.com/content/docs/DOC-4477*). The **Register Event Callback** function is used to obtain an "event" reference associated with the ".net" event source and a static reference to *speech check.vi*. This VI sets the **Value (Signaling)** property of a control referenced as "phrase."

The *speech test.vi* diagram in Figure 2.17 uses an **Event Structure** to monitor value changes in "phrase" if "speech?" is true. The "event"

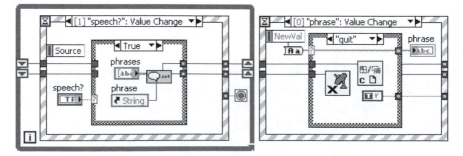

FIGURE 2.17   The *speech test.vi.*

FIGURE 2.18 The PowerPoint presentation control panel with buttons "open?," "first?," "previous?," "slide," "next?," "last?," and "end?" from left to right.

and ".net" references obtained from *speech setup.vi* shown are made available to all event cases using shift registers. Though these VIs worked well under Vista, deployment on an XP computer was unsuccessful.

CHALLENGE 9

Controlling a PowerPoint Presentation

PowerPoint is widely used to create and present slide shows. This Windows application can be controlled through ActiveX properties and methods. The challenge here is to build a simple control panel, like that shown in Figure 2.18, for opening, navigating, and closing a PowerPoint presentation. Admittedly, this control panel will not be of much use unless the PowerPoint application is installed on your computer.

SOLUTION 9

Controlling a PowerPoint Presentation

Creating an interactive application requires a suitable design pattern. Perhaps the simplest such pattern is the **User Interface Event Handler** shown in Figure 2.19. This pattern employs the **Event Structure** of the previous solution to handle user interactions. Shift registers make automation refnums available to all events. While this pattern is adequate for most user dialogs, a **Producer/Consumer** design pattern is often a much better choice for interactive applications and particularly advantageous for applications that invariably become more complicated with ongoing "feature" requests. This flexible design pattern is used in the database solution shown in Figure 2.26.

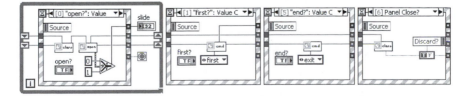

FIGURE 2.19 The *PowerPoint control panel.vi.*

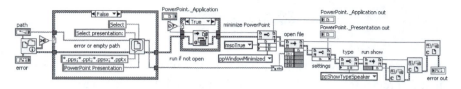

FIGURE 2.20   The *open PowerPoint presentation.vi.*

The *PowerPoint control panel.vi* employs three subVIs to open, navigate (cmd), and close a presentation through automation ref-nums for the application and the presentation (upper and lower shift registers). The PowerPoint application remains open until the user closes this VI. A filter event is used in the "Panel Close?" case to allow full operation without actually closing the window. This case would be converted to a notify event before building an application.

Navigating a presentation requires VIs to open a file as a slide show, move between slides, and close the file while possibly quitting PowerPoint. The *open PowerPoint presentation.vi* diagram in Figure 2.20 shows a method for opening a file at "path" in a minimized window and then starting a slide show. A file dialog appears if "path" is empty, and the application is opened only if not currently active. The user could also incorporate speech recognition from the previous solution to verbally navigate a slide show.

CHALLENGE 10

Reading and Writing Excel Worksheets

Excel workbooks are widely used to store and manipulate data. The Report Generation Toolkit allows LabVIEW applications to read and write specific worksheets within a workbook. Toolkit VIs dynamically call Excel subVIs typically located in *C:\Program Files\ National Instruments\LabVIEW 8.5\vi.lib\addons_office_exclsub. llb.* Building an application using toolkit VIs requires specification of dynamically called subVIs. Another approach is to bypass higher-level toolkit VIs and use the Excel subVIs directly to read and write worksheets. The challenge here is to perform these operations programmatically using this approach.

SOLUTION 10

Reading and Writing Excel Worksheets

Methods for reading and writing Excel worksheets are shown in Figures 2.21a and 2.21b. These VIs use ActiveX properties and

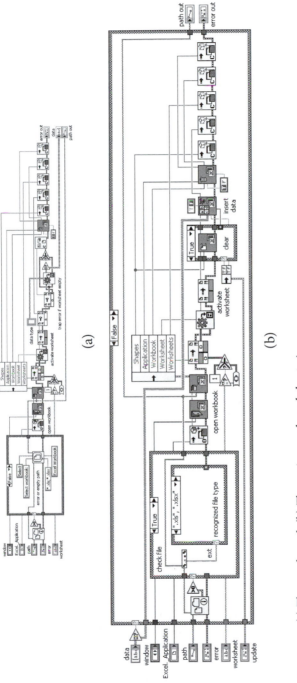

(a)

(b)

FIGURE 2.21 (a) The *read excel*. (b) The *write excel worksheet.vi* methods that read and write data as variants. Note the explicit conversion from variant to 2-D string array before the error trap in Figure 2.21a. The four automation refnums in addition to "Excel.Application" suggest the complexity of the ActiveX interface to Microsoft Office applications. The properties and methods available for these refnums can be explored by right-clicking on a wire and going to the submenu **Create>Properties...>** or **Create>Methods...>**. Clearly a Buck Rogers Decoder Ring or some other form of documentation is needed to make sense of such extensive options. These VIs have been tested using Excel 2003 workbooks (.xls) under Windows XP and Excel 2007 workbooks (.xlsx) under Windows Vista.

Figure 2.21a, the *read excel worksheet.vi* diagram, shows a method for reading all data from a specified "worksheet" in an Excel workbook at "path" using three toolkit subVIs. A file dialog is opened if "path" is empty, with the selected path output as "path out." The first worksheet in the specified workbook is used if "worksheet" is empty. Reading an empty worksheet does not result in an error.

The *write excel worksheet.vi* would typically use *Excel_Insert_Table.vi* from the Report Generation Toolkit to insert data into an Excel worksheet. However, this toolkit subVI formats all data using the general column format, which can be problematic for some strings. For example, the hex-like string "0102" will appear as the number 102 using the general cell format in Excel. Literal strings can be preserved by modifying the toolkit subVI to allow text format for all input data. The resulting *insert excel data.vi* adds a Boolean control "text?" that toggles between formats. This control and an option to replace worksheet contents are included in the "update" cluster found on *write excel worksheet.vi* diagram shown in Figure 2.21b.

CHALLENGE 11

Saving Panel Images Using Word

Microsoft Word is widely used to create documents on PCs. The Report Generation Toolkit allows LabVIEW applications to insert text and images into a Word document. This suggests that panel images for running applications could be programmatically saved to file. The challenge here is to save VI panels to a new or existing document using Word subVIs typically located in *C:\Program Files\ National Instruments\LabVIEW 8.5\vi.lib\addons_office_wordsub.*

llb. This approach is similar to the one used in the previous solution involving Excel subVIs.

SOLUTION 11

Saving Panel Images Using Word

A method for inserting text and images in Word and HTML documents is shown in Figure 2.22. ActiveX properties and methods used by Word are largely hidden in eight Report Generation Toolkit subVIs. The custom subVI *check file path. vi* is used to open a file dialog if "path" is empty. Always connect values to property and method terminals to avoid runtime errors. The *save panel image. vi* has been tested using Word 2003 files (.doc) under Windows XP and Word 2007 files (.docx) under Windows Vista.

Note that panel images must be saved to a temporary file before placing in a Word document. Each pane color is set to white prior to obtaining the JPEG panel image if "white?" is true (visible case). This image is stored in a tmp file and then deleted after saving the document. The *set page orientation. vi* is a custom subVI used to set page orientation (portrait/landscape) based on the aspect ratio of the panel image. You might want to remove this subVI if changing the page orientation creates a problem. In that case, you will have to use LabVIEW's **Write JPEG File.vi** as shown in the hidden True case for existing Word files.

CHALLENGE 12

Reading and Writing Access Database Tables

Database files are useful repositories of processed data and lookup tables. Such files are created using database management applications such as Microsoft Access. This application creates tables with fields

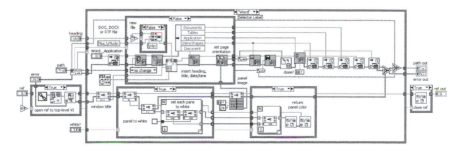

FIGURE 2.22 The *save panel image.vi.*

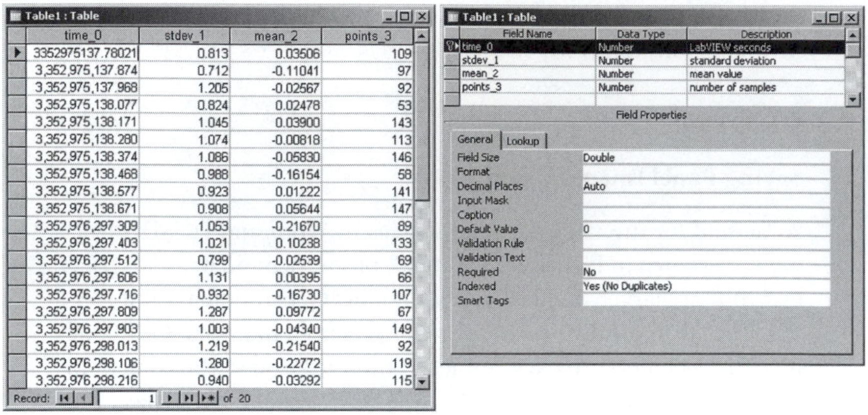

FIGURE 2.23   The Access table named "Table1."

in columns and records in rows as shown by the **Datasheet View** in Figure 2.23. Records as well as layout information shown in the **Design View** are saved to the Access 2003 file *stat.mdb*. The challenge here is to build a simple LabVIEW application for adding, clearing, selecting, and sorting records from this table.

The Access table named "Table1" is shown in **Datasheet View** and **Design View** in Figure 2.23. This table currently has 20 records consisting of time and statistics for sample data from a normally distributed population with zero mean and unit standard deviation. The field names and data types, shown in the **Design View,** will be needed to query this table by name. All fields shown are double-precision, floating-point numbers except "points_3," which is a signed 32-bit integer.

SOLUTION 12

Reading and Writing Access Tables

Most of the information needed by LabVIEW to work with a database file is encapsulated in a Microsoft data link (*.udl*) file. The **Data Link Properties** dialog shown in Figure 2.24 is obtained by double-clicking on file. A dialog can also be opened by selecting **Create Data Link...** from the **Tools** menu in LabVIEW or by right-clicking on a destination in Windows XP and selecting **New>Microsoft Data Link**. In either case, a service **Provider** appropriate to the database file specified on the **Connection** tab must be selected. The *.udl* file is typically placed in the same directory as the top-level VI. A

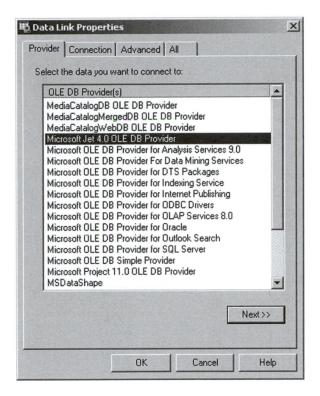

FIGURE 2.24   The Windows Data Link Properties dialog box.

connection to the database file will have to be manually established unless this file is in a fixed disk location.

LabVIEW interacts with Access *.mdb* files using the Microsoft Jet Database Engine. This 32-bit software has been included in every recent version of the operating system from Windows 2000 to Windows 7. The software is being replaced by the Microsoft Access Database Engine included with installations of Access 2007. This updated engine is backward compatible with Jet and additionally supports *.accdb* database files generated by newer versions of Access. Apparently, a 64-bit version of this software will be included with Microsoft Office 2010.

An application for interacting with "Table1" in the *stat.mdb* database file is shown in Figures 2.25 and 2.26. This application is based on the **Producer-Consumer** design pattern. Basically, a named queue is used to send "cmd" elements produced by an **Event Structure** to a separate command loop, which consumes these instructions by sequentially processing elements on the queue. Typically, all required

time	std dev	mean	points
2010/04/01 07:58:57	0.8135	0.03506	109
2010/04/01 07:58:57	0.7117	-0.1104	97
2010/04/01 07:58:57	1.205	-0.02567	92
2010/04/01 07:58:58	0.8239	0.02478	53
2010/04/01 07:58:58	1.045	0.03900	143
2010/04/01 07:58:58	1.074	-0.008183	113
2010/04/01 07:58:58	1.086	-0.05830	146
2010/04/01 07:58:58	0.9876	-0.1615	58
2010/04/01 07:58:58	0.9226	0.01222	141
2010/04/01 07:58:58	0.9079	0.05644	147
2010/04/01 08:18:17	1.053	-0.2167	89
2010/04/01 08:18:17	1.021	0.1024	133
2010/04/01 08:18:17	0.7985	-0.02539	69
2010/04/01 08:18:17	1.131	0.003949	66
2010/04/01 08:18:17	0.9315	-0.1673	107
2010/04/01 08:18:17	1.287	0.09772	67
2010/04/01 08:18:17	1.003	-0.04340	149
2010/04/01 08:18:18	1.219	-0.2154	92
2010/04/01 08:18:18	1.280	-0.2277	119
2010/04/01 08:18:18	0.9401	-0.03292	115

FIGURE 2.25 The *stat database viewer.vi* front panel.

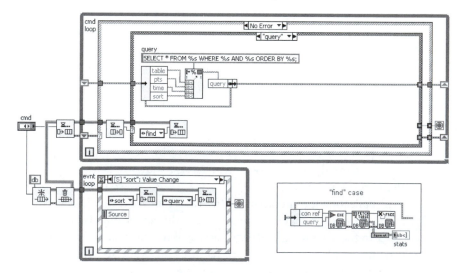

FIGURE 2.26 The *stat database viewer.vi*.

information is stored in a settings cluster that is shared between cases in the command loop.

The *stat database viewer.vi* panel for adding, clearing, selecting, and sorting records in "Table1" of the *stat.mdb* database file is shown in Figure 2.25. Controls specify a date/time range, sort field, and sample size constraint. Buttons are used to "add" records or "clear" the database file table. Results are displayed in the "stats" table indicator.

The *stat database viewer.vi* diagram in Figure 2.26 shows an example of the **Producer-Consumer** design pattern. Changing the "sort" menu on the panel causes an update of the "sort" string in the settings cluster. The subsequent "query" case in the "cmd loop" updates the "query" string prior to execution of the "find" case shown as an inset. Database Connectivity Toolkit VIs are used in this case to fetch a portion of "Table1" specified by the "query." This 2-D string array is formatted by the *format stats.vi* subVI prior to display in the "stats" table.

Errors are trapped and processed as they appear in the command loop. Errors generated in the event loop are typically associated with queue operations. While such errors are terminal, much like the movie *Final Destination,* they are initially trapped on entry to the command loop. Terminating this program from the command loop requires a hidden Boolean variable "exit?" to stop the event loop using the **Value (Signaling)** property. Other approaches to error handling and loop interaction are readily available as most LabVIEW developers have their own design philosophy.

In Figure 2.27, the *insert stats into table.vi* diagram shows a method for adding "stats" to a specified "table" in a database associated with connection reference "con ref." This process involves updating the "parameters" array with field values and specifying a query string. Note that this string has the same number of question marks as fields in the cluster array. This approach bypasses Database Connectivity Toolkit VIs used to construct the parameterized query based on an

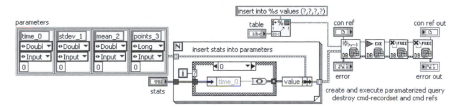

FIGURE 2.27 The *insert stats into table.vi.*

input cluster of record data. While these toolkit VIs can be used, the method shown in Figure 2.27 is faster for repeatedly adding records through an open database connection.

ACKNOWLEDGMENTS

The author benefited from reviews by Steve Solga (*http://electromecha. com*) and Matt Whitlock (*http://phaeron.net*). I greatly appreciate their time. I also benefited from feedback by other members of the ALARM user group (*http:/decibel.ni.com/content/groups/alarm*) at related presentations over the last few years. I'd also like to thank my coworkers Charlie Camacho, Bruce Carlson, Tony DeZarn, and Greg Lowry for helping me to expand my knowledge of LabVIEW and Windows.

RESOURCES

LabVIEW development system: Base/Full/Professional/Developer Suite for Windows (*http://www.ni.com/labview/*)

LabVIEW toolkits: Database Connectivity, Internet, Report Generation and VI Analyzer (*http://sine.ni.com/nips/cds/view/p/lang/en/nid/10447*)

Microsoft command reference: *WinCmdRef.chm* (*http://www.microsoft. com/downloads/details.aspx?familyid=5fb255ff-72da-4b08-a504-1b10266cf72a&displaylang=en*)

Microsoft .NET Framework: 4.0, 3.5 SP1, 3.0 (*http://msdn.microsoft. com/en-us/netframework/aa569263.aspx*)

Microsoft Office: Word, Excel, Access, PowerPoint (*http://office.microsoft. com*)

Microsoft SAPI 5.1: *SpeechSDK51.exe* (*http://www.microsoft.com/downloads/details.aspx?FamilyID=5e86ec97-40a7-453f-b0ee-6583171b4530&DisplayLang=en*)

Microsoft Windows: WordPad, Notepad, Event Viewer (*http://www. microsoft.com/windows/products/default.aspx*)

Optimumx: *shortcut.exe* (*http://optimumx.com/download/#Shortcut*)

PuTTY: *plink.exe* (*http://www.chiark.greenend.org.uk/~sgtatham/putty/download.html*)

TrueCrypt: *truecrypt.exe* (*http://www.truecrypt.org/downloads*)

WinZip: WinZip (*http://www.winzip.com/downwz.htm*) and *WZUNZIP. exe* (*http://www.winzip.com/downcl.htm*)

REFERENCE

Conway, Jon and Steve Watts. (2003). *A Software Engineering Approach to LabVIEW*. Upper Saddle River, NJ: Prentice Hall.

CHAPTER 3

A General Architecture for Image Processing

Craig Moore

CONTENTS

INTRODUCTION

This chapter assumes that the reader has a good working knowledge of LabVIEW, at least a basic understanding of the subjects discussed in the IMAQ Vision Concepts Manual, and experience developing small to medium applications. The chapter describes a framework for a general, image-based, particle analysis and characterization application known as the Bjorksten Particle Analysis System. With the architecture and tools employed, the reader is presented with a path to extend and add functionality to the application.

BACKGROUND

The example application provided in the chapter focuses on image metrology with special attention paid to a subset of digital image processing commonly known as particle analysis. Particle analysis in image processing is simply qualifying and quantifying or characterizing features of interest in an image. For these features to qualify as a particle, they must be separable from the background of the image. Particles are typically separated from the image background by contrast techniques such as threshold detection or edge detection. By defining the boundaries of these particles, they can then be characterized. Some of the more common characterization criteria are location, total area, roundness, center of mass, and orientation. The IMAQ Vision Development Module has over 80 of these characterization parameters.

Another image metrology technique for characterizing features and patterns in an image is called edge detection. Edge detection is used to find the boundaries of entire particles as described above. It is also used to find the beginning and ending edges of geometric shapes thus providing distance measurements. This technique is commonly used to analyze critical dimensions in semiconductors and production quality control.

CHALLENGE

The use of National Instrument's Vision Assistant and LV image processing tools allows the LabVIEW developer to quickly arrive at solutions to complex image processing problems. The challenge resides in identifying, creating, and integrating the various subVIs and tools to handle a series of individual problems and analyses. A solution is presented to demonstrate an example of this integration into an architecture to achieve code modularity that is adaptable and expandable, allowing the addition of third-party motion, motorized optics, and data acquisition.

SOLUTION

This solution focuses on particle analysis; however, the reader may wish to augment or replace this functionality with something like pattern recognition/measurement, color analysis, or data acquisition triggered image capture. The architecture is designed to be modular so that these things can be accomplished without much difficulty.

Although motion control was previously integrated in this solution, a system is presented that focuses on the core functionality—a complete solution for interfacing USB cameras and image processing related to particle analysis.

CONFIGURING THE SYSTEM

The application framework requires the following National Instruments development tools: LabVIEW 2009 or later, NI Vision Development Module 2009, and NI-IMAQdx 3.4 or later. If the reader is installing any of these tools for the first time, it is recommended to install in the order listed previously. These tools are available for download on the National Instruments website. Both LabVIEW and the NI Vision Development Module allow installation as trial versions, and NI-IMAQdx is available at no charge.

The only hardware required is a USB DirectShow imaging device. This includes most USB webcams, USB microscopes, and many other imaging products. The reader will need to install any drivers necessary to control the chosen DirectShow USB device. The operation of the device should be verified by accessing, configuring, and testing with the software from the device manufacturer prior to control with this application.

Although the framework source code included with this chapter has been developed and tested on Windows XP, it should run on all Windows XP and later systems. The source code package needs to be copied to the reader's chosen development directory. The reader may then launch the project file "Bjorksten Particle Analysis .lvproj" with LabVIEW.

USING THE APPLICATION: STEP-BY-STEP DEMONSTRATION SESSION

Video Source Configuration and Samples for Analysis

The video source is an inexpensive USB Microscope found on Amazon for around US$70. It has two magnifications, 20x and 400x. The samples used are confectionery decorations. These were finally settled upon as good examples because they demonstrate most of the parameters used in this particle analysis application—and they're delicious.

At this time, the USB imaging device should be connected and verified functional. Next, the main Bjorksten Particle Analysis System VI, "BPAS.vi," should be started (Figure 3.1). Note the names of the various areas and controls on the UI image as these names will be referenced

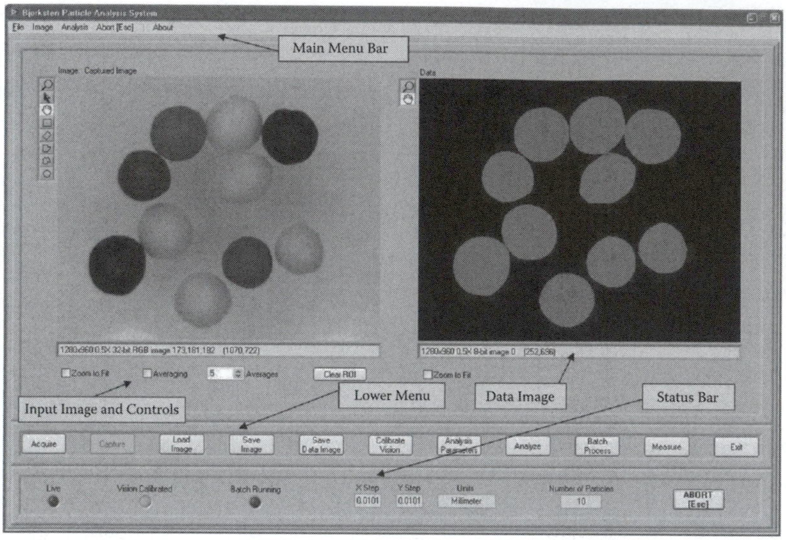

FIGURE 3.1   User interface.

throughout this text. From BPAS.vi Main Menu Bar, the Image: Setup Camera control is selected.

If a USB imaging device is detected, a dialog will appear as shown in Figure 3.2. The device to control should be selected from the Select Camera control and the video mode to use from the Select Video Mode control. The reader should note that the video mode control updates

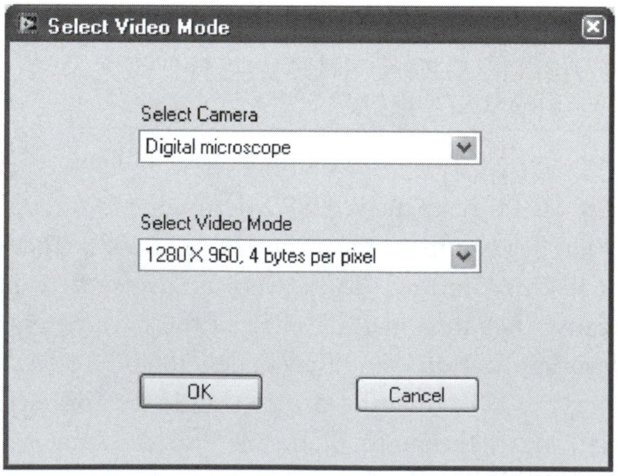

FIGURE 3.2   Selecting the video device and mode.

when the camera selection changes. Pressing OK accepts the changes and closes the dialog.

To begin acquiring images, the Acquire control needs to be selected from the Lower Menu. If everything is set up properly, live video will appear in the input image indicator and the Live LED indicator in the Status Bar will be on. If video frame averaging is desired, select the Averaging checkbox. The averaging algorithm performs a cumulative average of the number of frames shown in the Averages control.

Calibration

Before a particle analysis can be performed, it is necessary to calibrate. Calibration requires an image to be acquired that contains features of known length. For this example, an image of a millimeter scale was captured by selecting Calibrate Vision from the Lower Menu and Calibrate from Image in the ring control at the top of the dialog that appears. The image of the scale and the Select Calibration Type dialog are shown in Figure 3.3. Note that if the units/pixel value is known, the Manual Calibration selection may be used to enter the parameters manually.

After selecting Calibrate from Image and OK, a line calibration dialog is presented. The line tool is used to draw a line between two points of

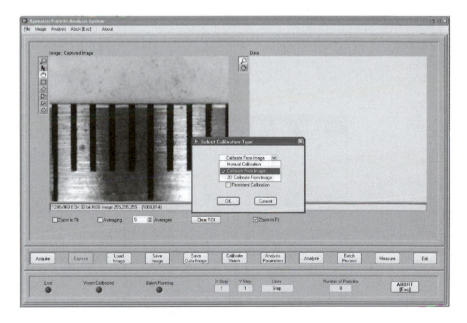

FIGURE 3.3   Selecting the calibration type.

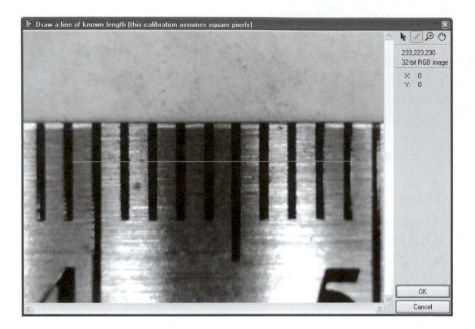

FIGURE 3.4   Line calibration dialog with a user drawn 10-mm line.

known distance as demonstrated in Figure 3.4. After drawing the line and selecting OK, enter the line length and units as shown in Figure 3.5.

If the imaging hardware does not have square pixels, the reader may use 2-D Calibrate from Image or Manual Calibration from the Calibrate Vision dialog.

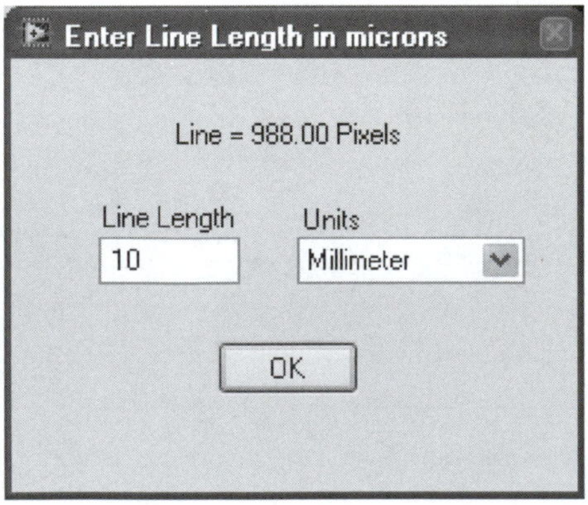

FIGURE 3.5   Applying the calibration to millimeters.

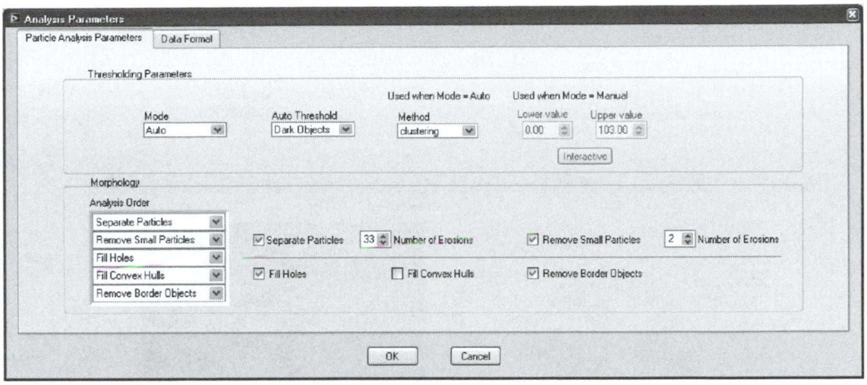

FIGURE 3.6   Parameters used to detect and process particles.

Analysis Setup

The analysis parameters must be set to values appropriate for the sample being analyzed. Select Analysis Parameters from the Lower Menu. Since high contrast is necessary for accurate particle analysis, the sample in this example is backlit to provide very dark objects on a white background. This allows the use of the Auto Clustering threshold method with good results. Note the analysis parameters used for this example and displayed in Figure 3.6.

If in some cases the auto threshold methods do not yield good results, the Manual Threshold mode may be used to isolate features of interest. The Manual Threshold dialog shown in Figure 3.7 is accessed by selecting Manual from the Mode ring control and then selecting the Interactive button. Isolation of dark or bright objects may be accomplished interactively by adjusting the sliders. Figure 3.8 shows the Data Format tab view. Any number of 81 particle measurement parameters may be selected for the analysis results. These parameters are described in the IMAQ Vision Concepts Manual.

Analysis and Processing

Note the good contrast from the backlit sample in Figure 3.9. The analysis is performed by selecting Analyze from the Lower Menu and a few seconds later (or milliseconds if particles do not need a lot of separation erosions), the data image as seen in the right-hand side of Figure 3.9 will appear. This operation is immediately followed by the Analysis Results dialog as shown in Figure 3.10. Selecting Save will save this analysis, including header information, in a Comma Separated Values (CSV) file.

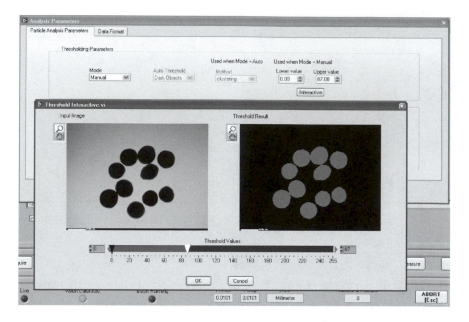

FIGURE 3.7   Using the manual threshold to isolate particles from the background.

Selecting Save Image or Save Data Image from the Lower Menu will save the respective image in one of several formats. A dialog will first be presented as shown in Figure 3.11 allowing the image file format selection. The NI Vision Development Module allows for embedding calibration information in PNG files, and in this system PNG format is always saved with the calibration information if the image has been calibrated. Additionally, options have been included to merge overlay graphics with the various formats. Choosing overlay graphics versions will save the units marker in the image.

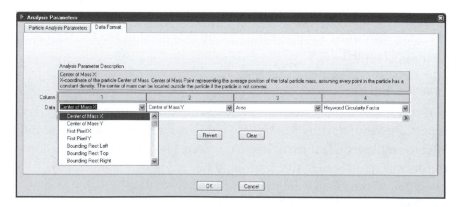

FIGURE 3.8   Selecting analysis results data to display and save.

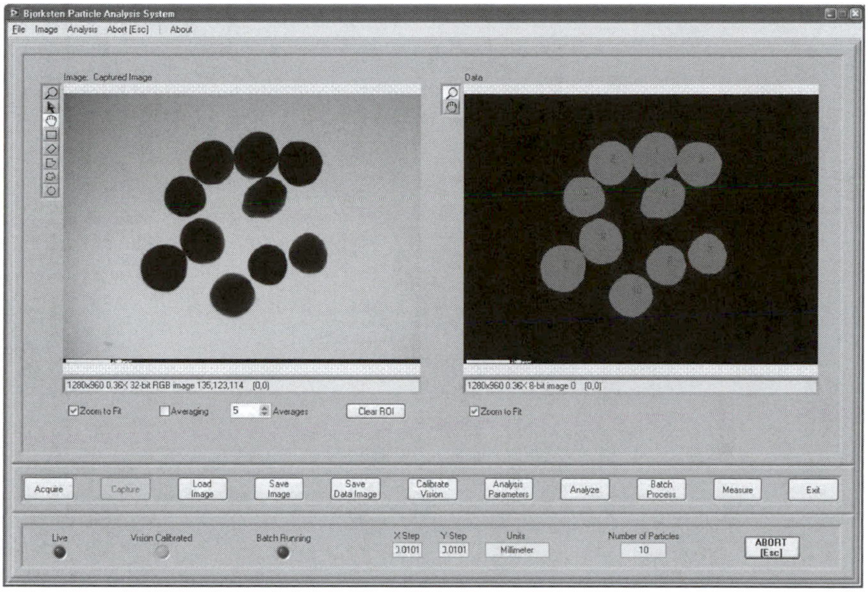

FIGURE 3.9   Analysis result with data image. Note that 10 were particles analyzed.

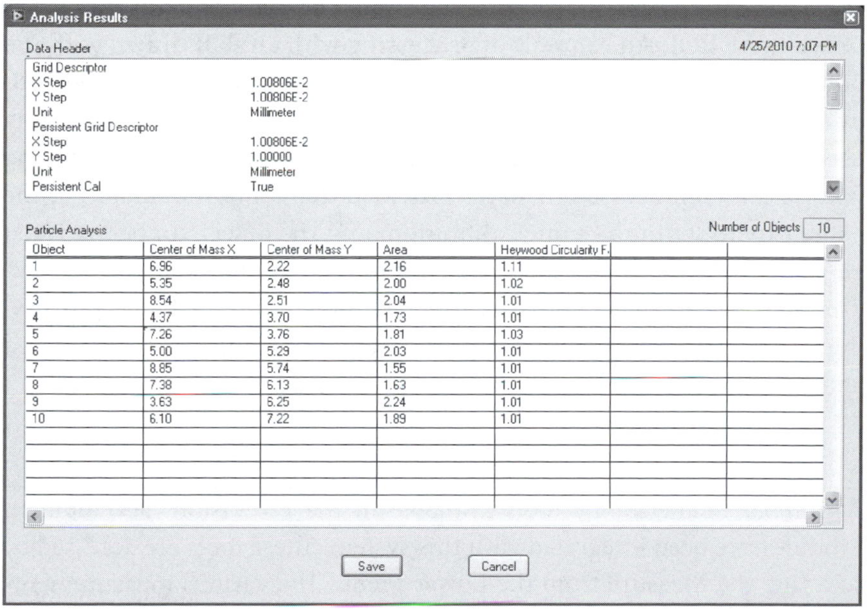

FIGURE 3.10   Analysis results for 10 particles.

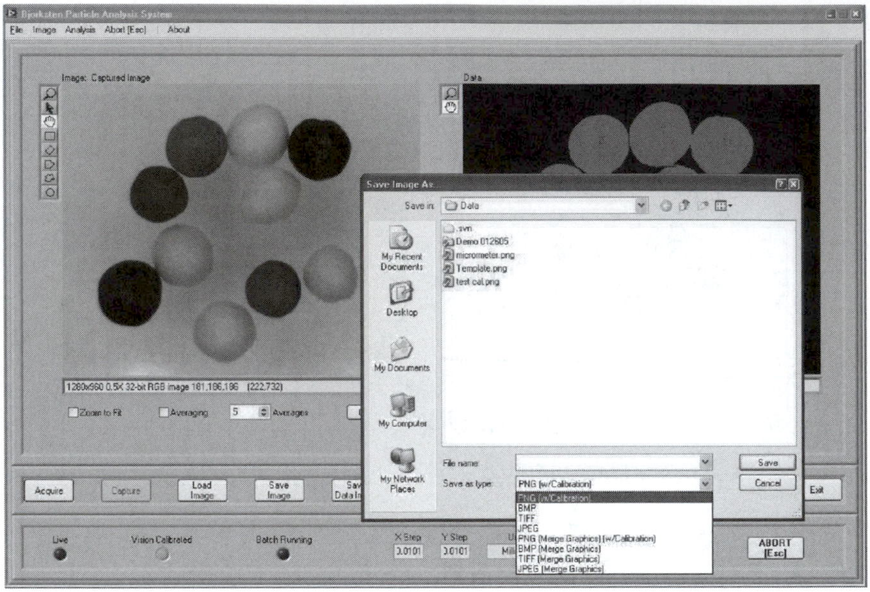

FIGURE 3.11   The Save Image dialog showing file type selection.

Note the Region of Interest (ROI) tools to the left of the input image. The application always processes images with respect to ROI. If the ROI is cleared (the Clear ROI button below the image control), the entire image is the ROI. An example of processing with an ROI drawn with the polygon tool is shown in Figure 3.12. Batches of images may be processed by selecting Batch Process from the Lower Menu. This mode is especially useful for processing images generated by another system. These existing images can be processed by BPAS provided that all of the images in the batch can be set to the same calibration, and the objects to be analyzed all share the same contrast characteristics (bright objects on dark background or vice versa). Note that this technique used for batch processing can be used for queuing up X, Y, Z stage coordinates for automated analysis of large samples.

METROLOGY TOOLS

A number of metrology tools available in the NI Vision Development Module have been integrated with this system. These tools are accessed by selecting the Measure from the Lower Menu. The various measurements are selected via the Measurements tab control. The reader should note that the vertical display of image tools to the left of the image changes to display

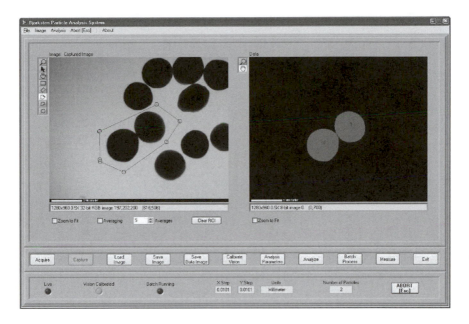

FIGURE 3.12   Example of applying a Region of Interest using the polygon tool from the palette.

tools appropriate for the measurement selected. Figures 3.13 through 3.18 demonstrate some of the measurements available.

THE CODE

Given that this text is targeted to LabVIEW developers, there will not be great detail about the code and coding style. Instead, this section describes the architecture and some of the utilities used to make this application work.

The project is organized as shown in Figure 3.19 with the project files, main vi, runtime menu file, and a config (cfg) file in the project root directory. Note that the config file was generated by running the application. The config file is a binary file that stores important user selected parameters, such as analysis and video settings, and should not be edited.

The directories are organized as follows:

Builds: contains a built executable, distribution files, and installer.

DLLs: contains SaveAs and Directory Browser utility VIs and the LVfileUtil DLL that they require.

FileIO: contains VIs for loading calibrated images and data file output.

SubVIs: contains all of the code related to image processing.

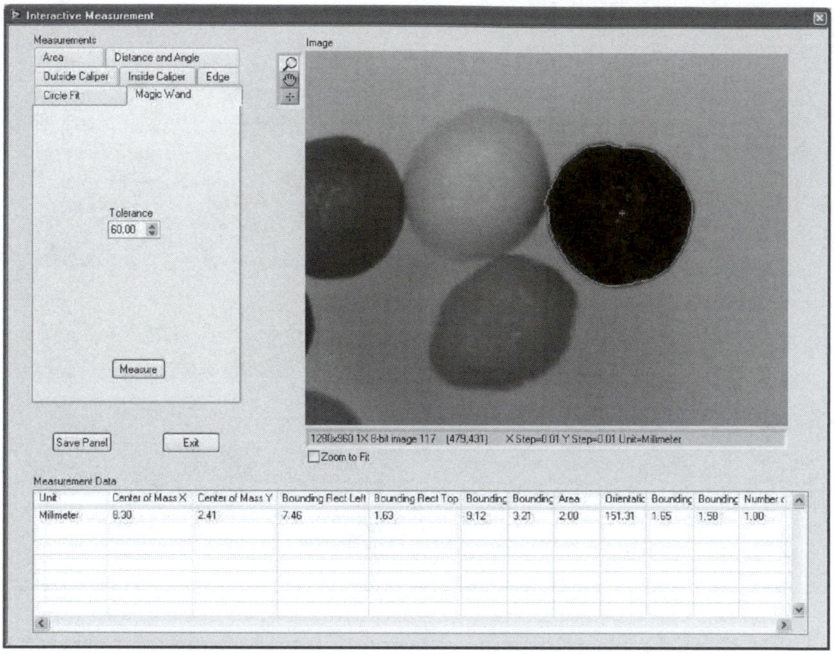

FIGURE 3.13   Example of applying the Magic Wand tool.

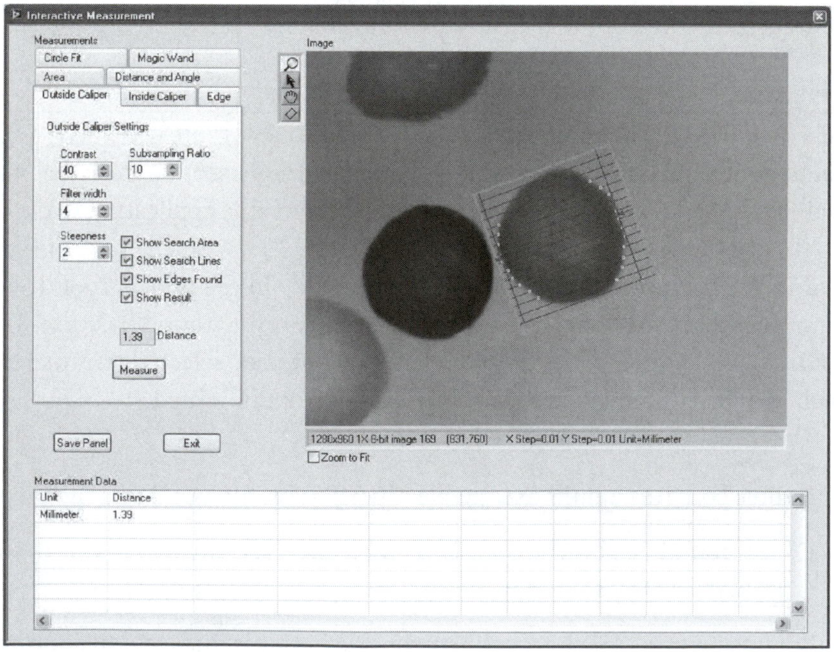

FIGURE 3.14   Example of the Outside Caliper measurement.

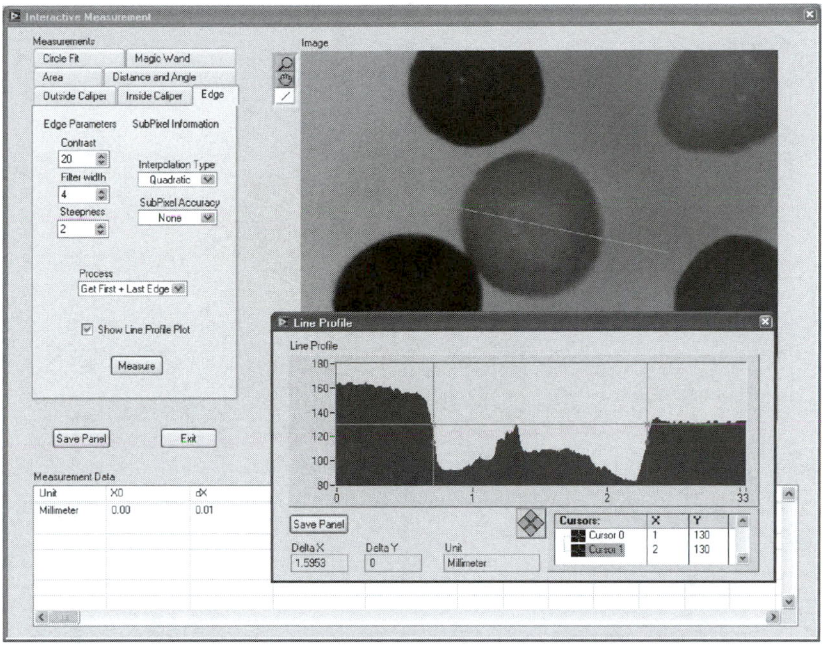

FIGURE 3.15   Example of the Edge tool showing a line profile graph.

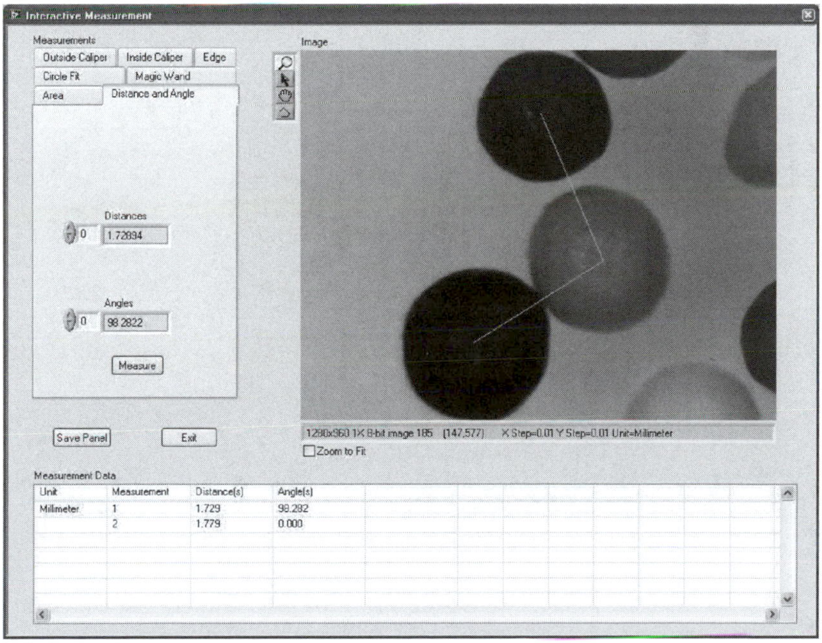

FIGURE 3.16   Example of measuring distance and angle.

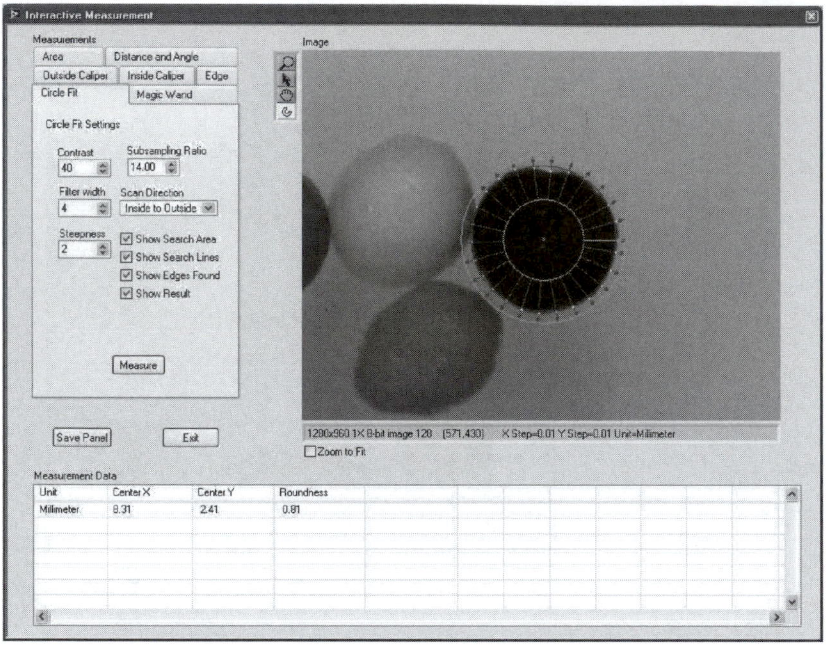

FIGURE 3.17   Example of applying the Circle Fit tool.

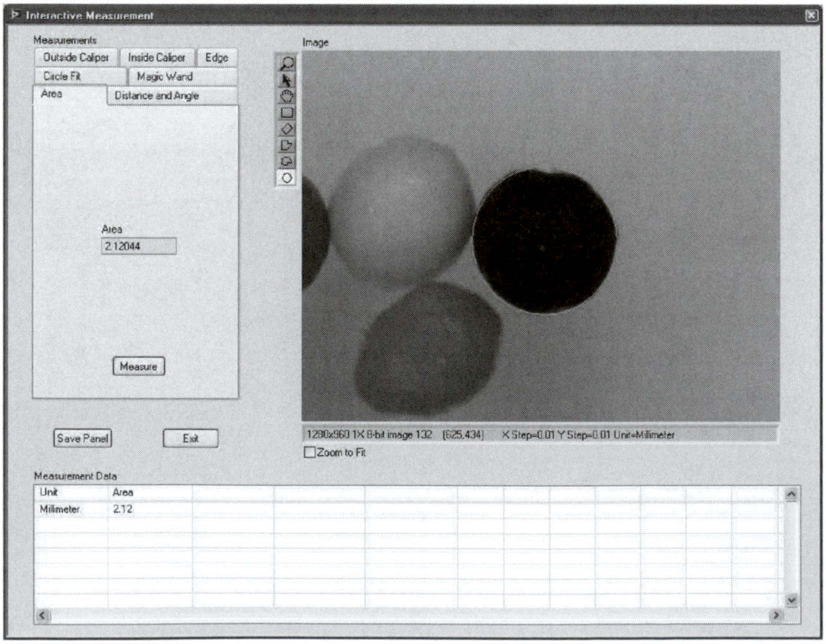

FIGURE 3.18   Example of applying the Area tool to measure a circular region.

FIGURE 3.19 Project directory structure.

TypeDefs: contains all of the data types used for the state machines and image processing parameters.

Utilities: contains general reusable code.

The architecture is a queue-based state machine driven by events and is shown in Figure 3.20. The UI loop hosts an event structure shown in the default idle state. Events generated by the user are used to perform

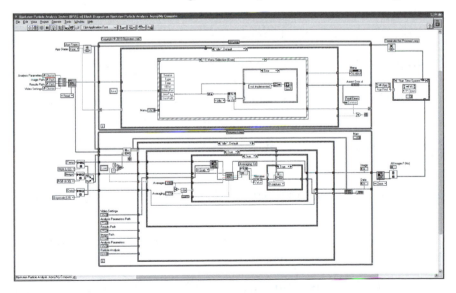

FIGURE 3.20 Main application block diagram showing UI and Process loops.

UI functions or dispatch one state defined by the App States Enum TypeDef (upper left of Figure 3.20). Most of the states in the UI state machine are used to load queue states to drive the Process Loop state machine. Other states are called to initialize, print, save front panel images, and exit. Both the UI Loop and Process Loop state machines use a subset of the App States TypeDef. The Process Loop state machine contains all of the states associated with acquiring, analyzing, and presenting images and data.

The following is a brief description of each state in the Process Loop state machine.

live: Initializes an IMAQ session and starts a grab acquisition. It enables the Live LED which is used as a status indicator to other states.

idle: When the Live LED is enabled, the Dequeue Element VI fires the idle state at 1ms timeouts. When Live is enabled, the idle state grabs a frame from the IMAQ device previously set up by the state "live." If averaging is enabled, frames are averaged for the number of frames specified in the front panel control, Averages. The state "capture" is called to stop the acquisition.

setup camera: Called when the Setup Camera control in the Upper Menu is selected. The current IMAQ USB session is closed and a camera and mode selection dialog is executed.

capture: Stops the current grab acquisition and, if calibrated, applies the current calibration parameters.

setup analysis: Runs the setup dialog (Figures 3.6 through 3.8). The parameters from this dialog are used to update the Analysis Parameters cluster. This cluster contains all of the information necessary to calibrate, process, and analyze images.

enable cal dialog: Helper state that resets calibration within the Analysis Parameters cluster.

calibrate: If no calibration is defined, the calibration dialog is executed (Figure 3.3).

apply calibration: Applies a calibration from the Analysis Parameters cluster to the current image.

threshold: Performs a threshold operation on the current image as defined in the Analysis Parameters cluster.

analyze: Performs a particle analysis as defined in the Analysis Parameters cluster.

save analysis: Allows the user to view and save data from a particle analysis. The header is constructed entirely from the Analysis Parameters cluster.

load image: Loads an image in any of the supported formats. PNG images can contain calibration information.

save image: Saves an image in any of the supported formats. PNG images can contain calibration information.

save data image: Saves an image in any of the supported formats. PNG images can contain calibration information.

measure: Calls the Measure dialog (Figures 3.13 through 3.18).

batch load: For each iteration of a batch, this state removes a path from the Batch Paths array control and then loads the image at that path for processing.

end batch: Resets the Batch Running. The main purpose for this state is to signal the UI Loop that the batch has finished.

abort: Called when the UI Loop enqueues this state at the opposite end causing it to be the very next state called. This is used to stop batch jobs but will stop any buffered queue operation within one state cycle.

show image window: Displays the current input image in an external display window (not used). Useful for displaying high resolution images in a larger screen space. Just add a control, event, and state in UI to try it out.

show data window: Displays the current data image in an external display window (not used). Useful for displaying high-resolution images in a larger screen space.

UTILITIES

The IMAQ USB Utility is located in a subdirectory in the Utilities directory. All the subVIs and controls necessary to implement the utility are in this subdirectory. A test module is included, "IMAQ USB Utility Test.vi."

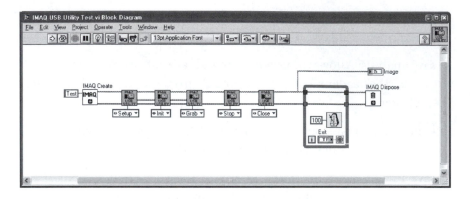

FIGURE 3.21   Block diagram showing a complete IMAQ utility test.

The block diagram is shown in Figure 3.21. This test VI illustrates the steps and order necessary to perform proper opening of an IMAQ session, image grab, and session close.

A couple of notable utilities are included. The "Build Header Table.vi" takes an input of an array of control references and outputs a spreadsheet ready 2-D array of strings containing a table of control label names and control values.

The "Read-Write FP Controls by Ref.vi" saves and restores control values from a file. The downside is that the file is binary and not human readable. The upside is that it is simple and handles any control type.

The "Config Filename from Top Level VI.vi" gets the top-level VI name and appends the string wired to Ext. Since this VI retrieves the top-level VI name from any level in the code, it is useful for randomly accessing a config file of the same base name as the main VI or application (in the case of a built executable).

The "Get Application Path.vi" returns the path to the directory containing the top-level VI or executable.

The two utilities in the DLL directory, "Win32 SaveAs Dialog.vi" and "Win32 Browse for Folder.vi," both access "LVfileUtil.DLL," which calls Win32API functions for SaveAs and Browse Folder dialogs. Since this application is Windows only (because we are using IMAQ), it was deemed sensible to have system-consistent dialogs for these functions. The LVfileUtil DLL source code is set up for the Borland C++ 5.5 compiler, which is available for free online.

CONCLUSION

This application is by no means a finished product, and it is my hope that the presented code will be expanded upon and shared online within the user community. The goal of this example was to provide developers with a tool that could be used directly out of the box; thus, functions that require other toolkits have been intentionally omitted. If the reader owns the Report Generation Toolkit, report output from the main application and the Measurement Utility may be added. It is hoped that readers enjoy exploring and extending this application as much as I enjoyed developing it.

RESOURCES

Embarcadero Developer Network

Borland C++ 5.5 (free download at edn.embarcadero.com)

National Instruments

LabVIEW 2009 or Higher

NI Vision Development Module 2009

NI-IMAQdx

IMAQ Vision Concepts Manual

CHAPTER 4

Radio Frequency Identification Read/ Write in Sports Science

Ian Fairweather

CONTENTS

INTRODUCTION

In fitness testing combines and related events, large numbers of athletes report to be evaluated and tested on a variety of physical performance test devices, instruments, and protocols. Due to the nature of the testing and the space requirements, tests are often spread over large areas of several buildings, tracks, fields, and gyms. This type of testing involves many operators, supervisors, and masses of athletes and often occurs over several days.

In this busy, noisy, and often hectic environment, operator errors often occur due to fatigue, noise, inexperience, stress, and extreme time pressures. These errors can lead to mistakes in the subsequently reported information and possibly test results being attributed to the wrong athlete. Further, some athletes can fail to satisfactorily complete a test or even front up to all of the required testing stations. These and other errors may not be detected before reports are being drafted, by which time athletes may have left the site. Errors of this kind cannot be tolerated in business or in elite levels of sport and can lead to embarrassment, frustration, loss of athlete selection, and, very importantly, lack of future testing business and loss of testing agency credibility. In this very intense environment, time is of the essence, and any methods that can be employed to help streamline the athlete testing processes are of great value.

CHALLENGE

The general challenge with this project was to enable the ready distribution of basic anthropometric information (e.g., height, weight, age) to the various and scattered testing stations that subsequently produce athlete performance data. It also required the timely collation of station athlete performance data and athlete check-in/check-out processes. The solution should be as free as possible from reporting errors while minimizing operator involvement and unnecessary time wasting.

The specific challenges of this project were to provide:

- A reliable method for easily identifying individual athletes

- A simple and reliable method of providing each testing station with basic athlete anthropometric measurements

- A reliable method of recording and collating the test station results for each athlete

- A reliable method of ensuring all athletes had completed all assigned tests

- Maximization of athlete throughput by minimizing station dwell time

- Minimization of the numbers of station operators and data notaries

- Minimization of the opportunity for human and data errors

- Minimization of system costs and setup time

SOLUTION OVERVIEW

At first blush, it would appear that a local area network (LAN)-based system could be used to communicate all information to and from all of the testing stations to a central computer; however, due to the nature of this type of testing, a variety of both indoor and outdoor venues at various locations have to be used. Most sport suitable venues have many sources of radio frequency interference (RFI) such as mercury vapor lighting that can easily affect the reliability of wireless-based LANs, particularly when signal strength is already compromised by distance or impeding structures. Further, many venues do not have existing or available networks (wired or wireless) that could be used for the purpose. Ensuring that connectivity would be constantly available to all stations at all times was therefore deemed to be a big and time-consuming task; therefore, a viable and cost-effective alternative was sought.

To solve these problems, a solution was developed whereby all testing stations involved in combine type testing would be fitted with a radio frequency identification (RFID) reader/writer. Upon their arrival at a combine check-in desk, athletes would be issued with a writable RFID card and lanyard. The card would be programmed with basic information such as the individual's name, age, address, team, and group. The athletes would then change into their performance gear and proceed with their card to the first testing stations where basic anthropometric data (e.g., height, weight, girth) would be added to their card, these measures being critical pieces of data required by subsequent testing stations.

Each testing station would include RFID-friendly LabVIEW-based programs that would not permit the test to begin until athletes had swiped their card, had been identified, and had their unique RFID number recorded on local hard drives. These programs would also read any anthropometric and other data that were required for that particular test protocol from the card and then acquire the performance data directly from the relevant instrumentation (at a few stations, where instrumentation was not appropriate for the test, the data would be entered by operator keyboard input). After having successfully completed the required test and the results having passed built in validation criteria, the information would be recorded directly into the athlete's RFID card (and to local hard drive for additional data security) before the athlete moved on to the next station. As athletes progress from station to station, additional data would then be added to the cards until the completion of all required tests.

Before athletes could leave the testing venue, they would be required to be "signed out." The sign-out procedure would require that their card was read by the check-out station reader where the associated computer would verify that all required tests had been satisfactorily completed. Upon successful completion, the collated content of athletes' cards would be saved to local disk.

Basic Combine/Systems Steps Summary

- Collate athlete attendance information.

- Athletes check-in and are issued with their RFID cards and lanyard.

- Basic information is written to card (e.g., name, group, team).

- Athletes present to anthropometric stations where information such as height and weight are added to card.

- Athletes present to a number of performance testing stations.

- Performance tests are run and validation criteria are checked.

- Performance data are written to card.

- Athlete checks-out and test results are collated and stored.

- Final analysis, report generation, and distribution of results to athletes and coaches.

BACKGROUND

RFID is an acronym for Radio Frequency Identification. As its name suggests, it employs low-frequency, low-power radio techniques to enable remote detection and identification of particular objects. The (passive tag)

RFID reader emits a radio frequency signal that is sensed by the RFID cards or tags and data embedded in the tags can be relayed via the RF field back to the reader and subsequently to the associated computer and software.

There is a large variety of RFID detection devices and a wide selection of tags on the market ranging from tiny passive tags requiring minimal space and operating over very small distances (centimeters) to active tags requiring battery power and capable of long distance communications (10s of meters or further).

Each RFID tag or card possesses a unique number called a unique item identifier (UID). This number is (usually) permanently embedded in the card at the time of manufacture and can't be readily altered. The UID is often eight or more Hex digits in length making many millions of unique IDs available. The UID can be readily linked in a database to, for example, a subject name, part number, or device serial number.

One of the most common uses of RFID is in libraries where passive tags are inserted into the spines of books. RFID readers at the counter can readily determine which book is being issued or returned by the received UID and the associated book title and can subsequently linked to the particular library customer. These devices are also commonly used as shop security mechanisms where alarms sound if goods are taken beyond the shop front without appropriate check-out procedures.

Some varieties of RFID tags and cards come supplied with internal memory capacity as well as the unique UID found in read-only types. This enables data to be stored directly in the cards rather than the UID being used as a reference pointer to relevant data stored in a remote database. Internal data storage capacity is now available in read/write cards from a few bytes to many kilobytes of user accessible memory.

RFID READER/WRITER DEVICES

Mifare (http://www.mifare.net) is a large manufacturer of RFID-related equipment. The Mifare RFID reader/writer devices chosen for this application are readily available from the Hong Kong-based supplier, the *RFID Shop* (http://www.rfidshop.com.hk/mf1.html). They are ISO14443A compatible, model MF1-RW-USB, 13.56 MHz high-frequency readers. These readers are capable of both reading the RFID card's unique UID as well as writing/reading limited amounts of data to and from their inbuilt nonvolatile memory (with the use of appropriate *writable* cards).

Connection to the associated PC is via a USB port or cable, which also
serves to power the device so no external power supply is required. The
device has very low power consumption, making it particularly useful for
outdoor laptop and netbook applications where mains power is not readily
available. Serial port versions of the reader are available, but these require
a separate mains-operated power source.

INSTALLING THE READER/WRITER

LabVIEW-based reader operation requires the installation of an associated DLL called *RR3036.dll,* which must be present in the working folder.
The DLL is supplied with the reader and can also be downloaded from the
RFID Shop website. LabVIEW VIs have been written using this DLL and
form the basis for this chapter. The drivers for this device are from Silicon
Labs (CP210x USB to Uart Bridge, Version 5.3.0.0), and the following files
are required to operate the device:

C:\Windows\System32\Drivers\silabenm.sys

C:\Windows\System32\Drivers\silabser.sys

C:\Windows\System32\WdfCoinstaller 01005.dll

These files are installed by the CP210xVCP_Win2K_XP_S2K3.exe file supplied as part of SDK from the RFID Shop.

Install the drivers, connect the reader, and open Windows Control
Panel>System>Hardware/Device Manager>Ports(Com & LPT) to expand
the view of the ports. Right-click on the "Silicon Labs CP210x USB to
UART Bridge (Com *n*). Select Properties>Port Settings>Advanced and
set Com Port Number = "Com5," the preferred port number for these
readers/drivers.

The MF1-RW-USB readers (Figure 4.1) are capable of reading/writing
data from their associated RFID cards from around 5–6 cm away (i.e.,
noncontact); however, in practice, the best or most reliable results are
achieved by placing the card directly on the reader's scanning surface.

The MF1-RW-USB reader supports only the ISO14443A RFID protocol.
The HT-TP-RW-D1 version of the reader, however, uses the same software
drivers and is capable of being software switched between ISO14443A,
ISO14443B, and ISO15693 protocols. Both readers look identical, function
similarly, and may be purchased for around the same price.

FIGURE 4.1   Mifare MF1-RW-USB reader.

RFID CARDS

Mifare also manufacture a wide range of RFID cards, *MIFARE Ultralight,
MIFARE Ultralight C, MIFARE Mini, MIFARE 1k, MIFARE 4k, MIFARE
Plus, MIFARE DesFire EV1,* and the *MIFARE SmartMX.* This chapter will
focus on two of the more popular and readily available cards containing
sufficient memory for most practical applications and available at a com-
petitive price: the MF1 IC S50 and the MF1 IC S70.

A credit-card-sized RFID (Figure 4.2) tag can be manufactured with
a small cutout at one end enabling the installation of a lanyard that can
be readily worn around the neck or wrist. This simple feature enables
improved user convenience and also provides better card (and data)
security.

FIGURE 4.2   Mifare S70 type card with lanyard attachment punch-out.

The cost of a single card in quantities of 500 is around US$1; the readers (complete with USB cable and driver) are around US$100 each. For this project where overall costs, simplicity, and ease of setup were viewed as paramount considerations, the devices chosen therefore seemed to fit the basic requirements.

Tags are available in a variety of styles, shapes, and sizes including wrist watch styles, cards, and doughnuts. Users should select the style that best suits their application and mounting/attachment needs.

It should be noted that the following sections are not intended to be a comprehensive and all-encompassing description of all ISO14443A functionality; however, sufficient information is provided to enable the user to develop a simple functional system for the reading and writing of data to and from RFID cards and tags. It is left to the reader to do more detailed research into the intricacies of the protocol if additional functionality is required.

ISO 14443A CARD MEMORY ORGANIZATION

The Mifare S50 card has 1 kB, and the S70 has 4 kB of user accessible memory. Card memory is arranged in a matrix of sectors and blocks, with the S50 version having 16 sectors and four blocks and the S70 having 32 sectors of four blocks plus an extra eight sectors containing 16 blocks each.

The first block (sector 0, block 0) of all cards contains the UID and card manufacturer code. This location cannot be written to. The last block in all sectors contains information pertaining to security access to that sector, and *unnecessary writes to this block should be avoided* unless great care is taken. This block contains a security Key A, a control section, and a security Key B. Since security was not an issue in this application, no attempt was made to implement access to this block. Further, writing inappropriate data to this block can permanently lock the sector, causing future writes to the whole sector to fail.

Memory organization in other card types varies, so users should carefully consult the relevant device specifications before attempting to write code to different card types. If developing code for other card types, expect to damage at least several cards during the development process. For more specific information on card memory usage in this application, refer to the "Memory Location.vi" described in the Application subVI section of this chapter.

In this application, there was a requirement for three different data types to be written to the card: (1) text strings, such as surname and street name;

(2) floating point (FP) numbers, such as athlete height and weight; and (3) a set of integers. ASCII text strings can be easily written to the 14443A cards. To accommodate all three data types needed and to simplify coding, a decision was taken to have all data stored on the cards as ASCII text strings. Floating point and integer numbers therefore are first converted to text strings before being saved to card memory. The capacity of 16 text characters per block of the ISO14443 cards was deemed sufficient for the purpose as most names, and addresses could be readily accommodated within that space. Having a 16-character width for the floating point data also proved to be satisfactory for the types of data that were being generated (i.e., Subject weight = +101.12345678901 kg, Shuttle run time = +15.123456789012 seconds) enabling more than enough resolution and range and could also accommodate a sign and % symbol if required. For application convenience, integers were treated as floating point numbers in this project. A more efficient or different memory usage process could be readily adopted for applications where larger numbers of variables or wider string widths are needed.

BASIC RFID OPERATIONS

Writing ASCII Strings

Writing ASCII text to the cards (Figure 4.3) involves the following process and order of execution:

1. Open com port (Auto Open Com Port.vi).

2. Get reader information, optional (Reader Info.vi).

3. Request (Request.vi).

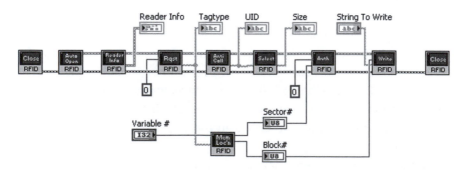

FIGURE 4.3   Procedure for writing strings to ISO14443A card.

4. Anti-collision (Anticoll.vi).

5. Select (Select.vi).

6. Determine the appropriate sector and block locale to write to (Memory Location.vi).

7. Authentication (Authentication.vi).

8. Write (Write String.vi).

9. Loop to #6 as required.

10. Close com port (Close Com Port.vi).

Once the serial port is open and the card UID has been identified, writing a 16-character string to the cards takes around 5 ms when executed in a LabVIEW loop. Text written to the cards using this method can have a maximum length of 16 characters. Text strings longer than 16 characters will be truncated to 16. If an original text string is being replaced by a new string shorter than the original, the residual characters in the old string will all be set to "".

Figure 4.3 demonstrates the basic sequencing required to perform an RFID card string write. The cluster joining most VIs (at the top) is designed to more easily pass on communications port and command address information to all relevant VIs. The communications port is shown to be closed and then reopened at the beginning of the VI. This was done in an effort to avoid annoying errors during software development and editing stages. If the port is already open and subsequent attempts to open it again are made, com port errors will occur. It was expedient therefore to always close the com port before making a call to open it.

Refer to the application high-level VIs section titled "Writing Text Strings" for further information and process simplification.

Reading ASCII Strings

Reading ASCII text from the cards (Figure 4.4) involves the following process and order of execution:

1. Open com port (Auto Open Com Port.vi).

2. Get reader information, optional (Reader Info.vi).

3. Request (Request.vi).

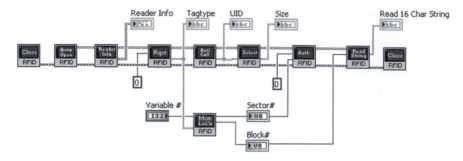

FIGURE 4.4   Procedure for reading strings from ISO14443A card.

4. Anti-collision (Anticoll.vi).

5. Determine the appropriate sector and block numbers to write to (Memory Location.vi).

6. Authentication (Authentication.vi).

7. Read (Read String.vi).

8. Loop to #5 as required.

9. Close com port (Close Com Port.vi).

Figure 4.4 shows the basic sequencing needed to perform a card string read. Once the serial port is open and the card's UID has been identified, reading a 16-character ASCII from a card takes around 5 ms when executed in a loop.

Application High-Level VIs

Writing Text Strings

The *Write String.vi* (Figure 4.5) first establishes that a card is in the field and then reads its UID and type using the "Info.vi." The VI then takes the variable location supplied via the "String Cluster" and computes the relevant block and sector number. The memory location is tested against the available card memory and if it is within the legal range, the "Variable Exists" Boolean is made True. The ASCII text supplied in the "String Cluster" is then written to the card memory. If the text string is > 16 characters in length the "String Truncated" Boolean will be made true; however, the first 16 characters will still be written to the card. The string that was written to the card is read and returned in the "Read String" indicator. If the write was successful the "Write OK" Boolean will be

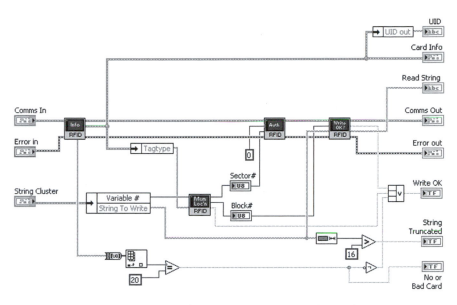

FIGURE 4.5   *Write String.vi* used to write text string to the card.

returned True. If no card is detected in the field, the Bad or "No Card Boolean" is returned True.

Reading Text Strings

This VI (Figure 4.6) first establishes that a card is in the field and then reads the UID and type using the *Info.vi*. The VI then takes the variable location passed to it and using the *Memory Location.vi* computes the appropriate block and sector numbers for the read. The memory location is tested against the available card memory and if it is within the legal range, the "Variable Exists" Boolean is made True. The memory location is then read and the ASCII text from that location is returned in the "Read String" indicator. If no card is detected in the field, the "Bad or No Card Boolean" is returned True.

Writing Floating Point Numbers

To write FP numbers to the cards using the *Write Float.vi* (Figure 4.7), the user supplies the card memory location and the floating point number to be saved in the "Data Cluster." As floating point numbers can't be written directly to the cards, all data should preferably be in ASCII text format before performing any read/write operations. The input FP number is converted (including its sign) to an ASCII string using the standard LV "Number to Fractional String" function. The memory location is passed to the *Memory*

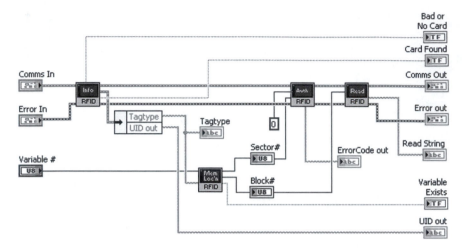

FIGURE 4.6 *Read String.vi* used to read a text string from a card.

Location.vi which computes (for the specific card type) a suitable sector and block number using the lookup table described earlier. The FP string is then saved to the card using the *Write & Verify.vi*, which also checks that the data written to the card matches the original input FP number.

The default settings for the FP to string conversion are Width=16 and Precision=6. These settings can be manipulated to best suit particular applications; however, the default settings were found in practice to be suitable for all physiological variables used.

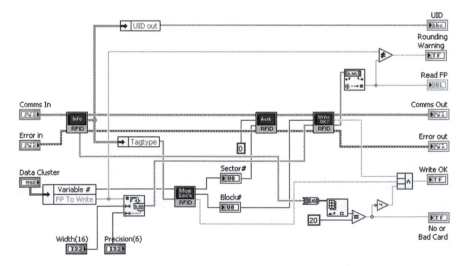

FIGURE 4.7 *Write Float.vi* used to write a single floating point number.

If the write process is verified, the "Write OK" Boolean will be returned True. If some rounding or truncation of the stored FP variable occurs due to lack of conversion precision, the Rounding Boolean will be made True. If no card is found in the RF field during the write process (20H being returned by the *Info.vi*), the "No or Bad Card" Boolean will be made True.

For this application, there were more than sufficient characters in a single 16-character ASCII text string to also allow very large integers to be stored and read using the floating point method previously described.

Read Floating Point Numbers

The *Read Float.vi* (Figure 4.8) first establishes that a card is in the RF field and reads the UID and card type using the *Info.vi*. The VI then takes the variable location passed to it in the Data Cluster and uses the *Memory Location.vi* to compute the appropriate block and sector numbers for the read. This memory location is also tested for validity with the card type being used, and if it is within range the "In Range" Boolean is made True. The memory location is then read and the ASCII text from that location is converted back into a floating point number using the *Fract/Exp String to Number* function.

The read number is also tested to see if it is indeed a number rather than a string. If it is a number, the "FP Num?" Boolean is made True. If no card is in the field, the "Bad or No Card Boolean" is returned True.

Application integers were read using the same floating point process.

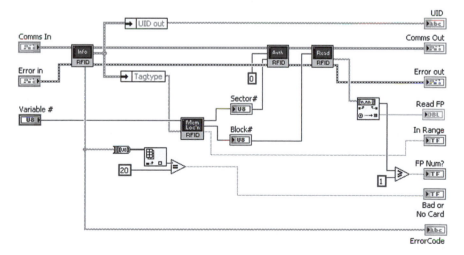

FIGURE 4.8   *Read Float.vi* reads a single floating point number from card.

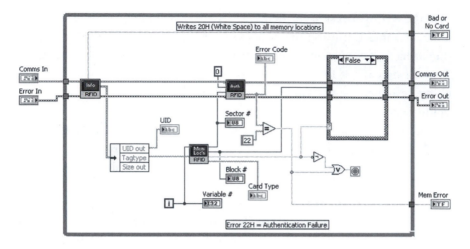

FIGURE 4.9   *Erase Card.vi.*

Erase Card

This VI can be used to erase information previously recorded on the card or to test that the card's available memory is accessible and fully functional before being distributed.

The *Erase Card.vi* VI (Figure 4.9) writes 20H to each variable location and subsequently reads the written value back for comparison. If the written data do not equal the read data, the VI aborts with an error and reports the sector and block number where the first error was encountered. Depending on the card being erased or tested (and once the UID has been read), the completed cycle will take around 4 seconds for the S50 cards and around 16 seconds for the S70 type cards (having about four times the memory capacity).

Read Whole Card

This VI (Figure 4.10) reads the entire content of a card and creates a <CR> delimited text string containing all ASCII text in the card's accessible memory (including both text and numeric data). If no card is detected in the field, the "Bad or No Card" and the "Mem Error" Booleans are both set to True. If the card is removed during the reading process, both Booleans (above) are set to True, but the data read up to that point will be accrued in the "String" indicator and the "# Read" indicator will contain the location at which the process was interrupted when the card was removed. Empty memory locations will be returned with 20H followed by 0DH.

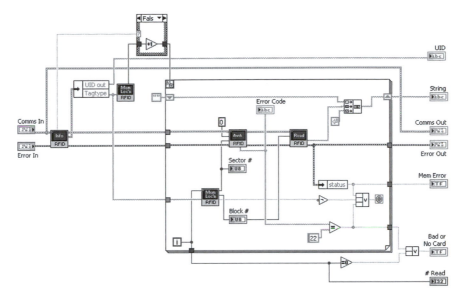

FIGURE 4.10   *Read Whole Card.vi.*

Read Card Section.vi

This VI (Figure 4.11) is similar to the *Read Whole Card.vi*; however, limits may be placed on the starting and ending points of the card read. This is done by setting the starting point of the read and the number of successive memory locations to be read using the "Start Location" and "# Locations" controls, respectively. <CR> delimited text strings will be accrued in the "String" indicator. If no card is detected in the field, the "Bad or No Card" and the "Mem Error" Booleans are both set to True. If the card is removed during the reading process both Booleans (above) are set to True, but the data read up to that point will be accrued in the "String" indicator and the "# Read" indicator will contain the location at which the process was interrupted.

Empty memory locations in the section read will be returned with 20H followed by 0DH.

RFID Function Master

This application required that both text and multiple floating point numbers were able to be easily and reliably written to the cards. To greatly simply the coding process for the multiple numbers of programs that were required to be written for the different athlete testing stations, a master VI was created. This master VI (*14443A Master.vi*, Figure 4.12) enables any

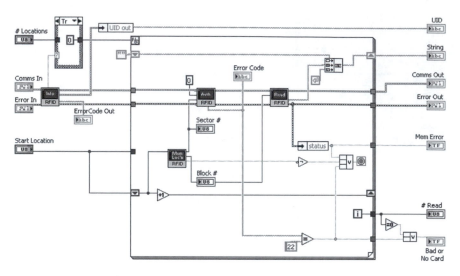

FIGURE 4.11   *Read Card Section.vi.*

one of the following functions to be selected and operated through the use/placement of a single VI on the calling program's wiring diagram:

Detect reader and open communication port

Get reader information

Get RFID card info (e.g., UID, card type)

Read a stored 16-character ASCII string

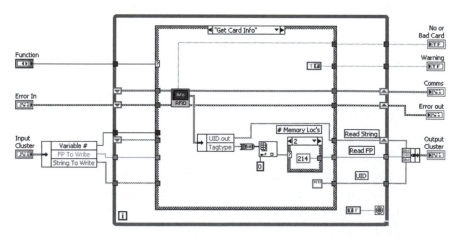

FIGURE 4.12   *14443A Master.vi* shown in the "Get Card Info" state.

Write a 16-character ASCII string to card

Read a stored floating point number

Write a floating point number to card

Read a section of card

Read the entire card content

Erase a card

Beep the reader buzzer

Flash the reader LED

Close com port

The required function (from the previous list) can be selected by passing the enum control "Function" containing the operation needed. During read and write operations, the "Input Cluster" is used to also provide the relevant data. In all read functions, the memory location of the variable is required in the "Input Cluster." During all write functions the memory location and variable data (numeric or text) are required to be supplied.

Note: To read, write, or obtain card information using this VI, call the *Auto Open Com Port.vi* function once. When the port is open and operative, further calls to the port when using these commands are no longer needed. Com port and command addresses are stored in the VI's uninitialized shift register and are passed to the subVI's functions automatically if and as required.

APPLICATION SUBVIs
Memory Location

To more easily access the card's memory using standard LabVIEW nomenclature, a lookup table was developed (Figure 4.13) to cover all available memory in both S50 and S70 cards. This 2-D array lookup table could then be used to provide the sector and block number of a particular stored variable by the use of a single array index or "variable#." A truncated version of the lookup table is provided showing the assignment of sector and block numbers.

The lookup table was developed in three parts corresponding to the three different sets of memory mapping required. S50 cards use only the S50 section of the table (Figure 4.13), whereas the S70 cards use both

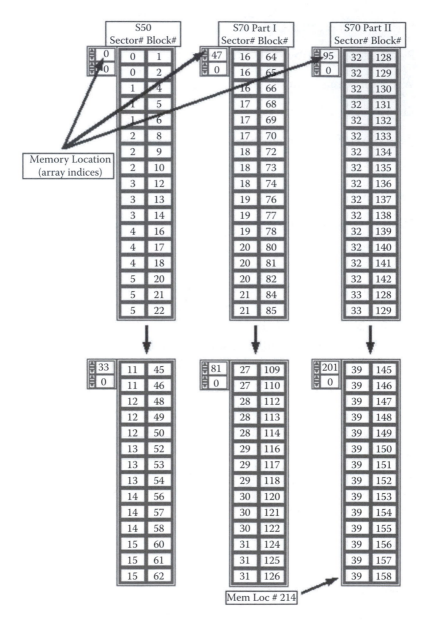

FIGURE 4.13   14443A lookup table.

the S50 section and parts I and II of the S70 section. S50 cards use only the lower 47 array indices (0–46) of the table; the S70 cards can use all 215 (0–214). In simple terms, the S50 cards have 46 × 16 character available memory locations, whereas the S70 have 214 × 16 character memory usable locations.

It is very important to note the missing block numbers in all sector sequences in the lookup table (the last block in every sector has been removed). Examples of this are sector 1 in which block 7 is missing and sector 17 in which block 71 is missing. The last block in each sector has been removed to avoid that sector being locked inadvertently by writing inappropriate data to it. Sector #0, Block #0 is precluded as this is UID location.

The VI called *Memory Location.vi* (Figure 4.14) uses the lookup table (Figure 4.13). It takes a user input "variable #" and determines (for the particular card type being used) a suitable sector and block number to be subsequently passed to the *Write String.vi*, *Write Numeric.vi*, or *Write Float.vi*. The lookup table carefully maps variables around block 0 and the last block in all sectors to avoid inadvertantly writing to security blocks. If these blocks are written to, accessing the blocks in the future may be precluded.

For applications that require larger amounts of numeric or string storage, more efficient use of the available memory could be made by encoding blocks with two or more variables contained within each. However, with more than 200 variables available in the S70 type cards, there was more than sufficient memory available for this application.

Card Information

To assist with simplifying the reading and writing coding processes, the request, anti-collision, and select functions have been condensed into one VI called *Info.vi* shown in Figure 4.15. The VI returns the card type, UID, and the card's memory size. The "Comms In" and "Comms Out" clusters

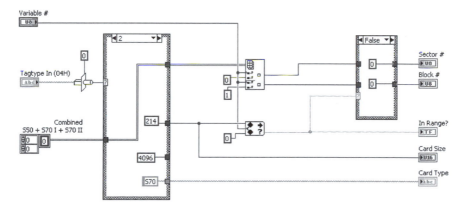

FIGURE 4.14 *Memory Location.vi*, used to determine block and sector numbers for the variable to be stored or read.

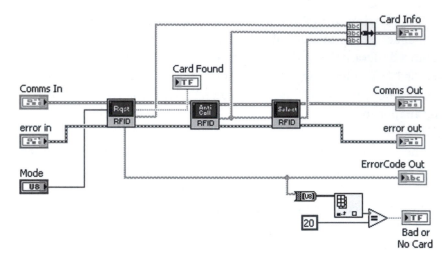

FIGURE 4.15   The *Info.vi*, used to get card specific information.

are used to simplify the carriage of communications port and command address information to and from most of the read/write related VIs in this chapter.

Write and Verify

This VI (Figure 4.16) has been used extensively in this application instead of the simple *Write String.vi* to ensure that the string data written to the card is actually what is subsequently retrieved from it. The VI simply performs a read immediately after the write and does a comparison of what was intended to be written with what was read. The VI assumes that ASCII strings will be written (i.e., no stored data will be < 30H).

Exchanging the *Write & Verify.vi* for the *Write.vi* adds an extra 5 ms to the cycle time, making it ~9 to 10 ms per variable when inside a loop

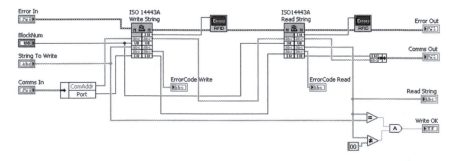

FIGURE 4.16   The *Write & Verify.vi*.

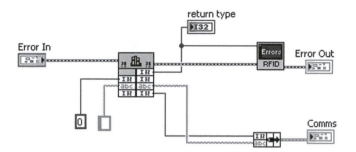

FIGURE 4.17 The *Auto Open Com Port.vi*.

instead of ~5. If cycle time isn't an important factor, the *Write & Verify.vi* alternative should be used to assure data integrity.

LOWER-LEVEL VIs

Open Communications Port

The *Auto Open Com Port.vi* (Figure 4.17) searches the communications ports and opens the port associated with the RFID reader if found. In Windows XP, Control Panel>System>Hardware>Device Manager>Ports, the port has the following text associated with it: "Silicon Labs CP210x USB to UART Bridge (COMn)" where n = port#.

Close Communications Port

The *Close Com Port.vi* (Figure 4.18) terminates a communications session and frees the port up for use by alternate devices. In practice, it is a good idea to place this VI before the *Auto Open Port.vi* when building applications as an error will be created if the port is already open.

RFID Reader Information

The *Get Reader Information.vi* (Figure 4.19) returns basic information about the reader connected to the communications port. The returned data include:

Firmware version: First byte is the version number, and second byte is the subversion number

FIGURE 4.18 The *Close Com Port.vi*.

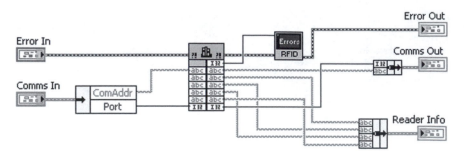

FIGURE 4.19   The *Get Reader Information.vi.*

Reader type: RR1φ36

TR type: Supported protocol

Inventory scan time: Time to scan UID, default is 300 ms

Set Beep

The reader has an inbuilt configurable buzzer. The DLL permits the user to set the beep duration time, quiet time, and the number of beeps. The *Set Beep.vi* (Figure 4.20) is used to sound the buzzer. The "Buzzer" cluster contains three configurable buzzer-related parameters of the reader:

Open time = Time the buzzer is sounding (0 to 255), in increments of 50 ms.

Close time = Time buzzer is silent (0 to 255), in increments of 50 ms.

Repeat count = Number of times the buzzer is sounded.

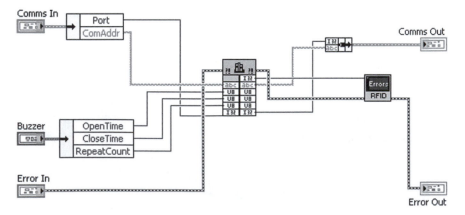

FIGURE 4.20   The *Set Beep.vi.*

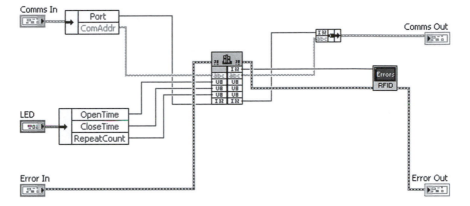

FIGURE 4.21 The *Set LED.vi.*

Set LED

The reader has an inbuilt configurable bicolor LED. The LED by default glows red when power is applied to the reader. The DLL permits the user to make the LED go green for a certain duration before returning to red and also permits the user to set the number of red to Green cycles to be repeated. The *Set LED.vi* (Figure 4.21) permits the LED color to programmatically be altered. The "LED" cluster contains three configurable LED-related parameters of the reader:

Open time = Time the LED is green (0 to 255), in increments of 50 ms.

Close time = Time the LED is red (0 to 255), in increments of 50 ms.

Repeat count = Number of times the LED changes color.

Read

This VI (Figure 4.22) reads the text string stored at the location "BlockNum" from ISO14443A tags and cards.

Write

This VI (Figure 4.23) writes the text string passed to it at the location "BlockNum" to a writable ISO14443A tag and card. No verification of the written data is performed. Refer to the *Write & Verify.vi.* A single write will take approximately 5 ms when performed in a loop.

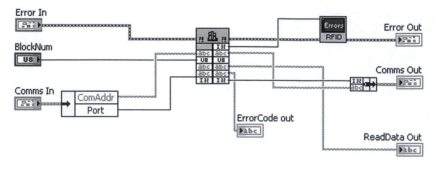

FIGURE 4.22   The *Read.vi* reads an ASCII text string from a single block.

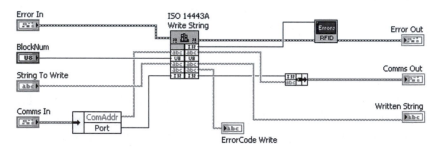

FIGURE 4.23   The *Write String.vi* writes a text string to a single block.

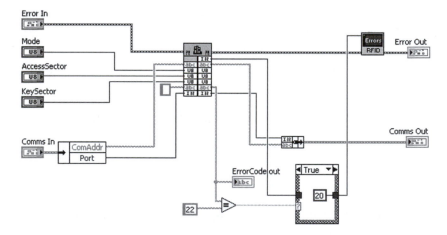

FIGURE 4.24   The *Authenticate.vi*.

Authentication

This is a security feature, providing a mechanism for mutual authentication between tag and reader/writer for a given sector or block. The tag's sector data can be read or written only following a successful authentication exchange (Figure 4.24).

Request

This VI is used to detect the presence of an ISO14443A tag in the RF field and returns the tag type if a tag responds (Figure 4.25):

Tag type = 04H, S50

Tag type = 02H, S70

Tag type = 44H, Ultralight

Anti-collision

This VI is used to return the unique UID of a single card should multiple cards be in the RF field (Figure 4.26).

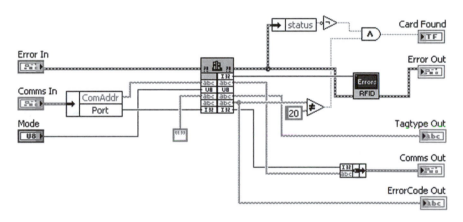

FIGURE 4.25   The *Request.vi.*

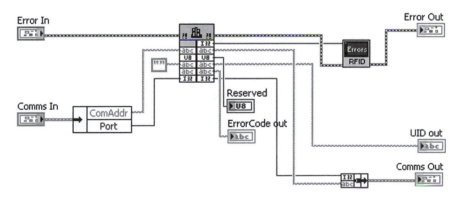

FIGURE 4.26   The *Anticoll.vi.*

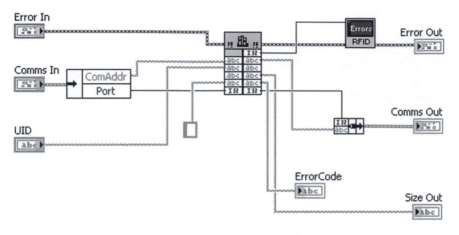

FIGURE 4.27 The *Select.vi*.

Select

This function selects the specified tag as the preferred one to be communicated with (Figure 4.27).

Switch to ISO14443A Mode

The *14443A Mode.vi* is used or required only if operating the multimode type readers such as the HT-TP-RW-D1. It is used to switch the reader into the ISO14443A mode from any other current mode. For single-mode readers, this VI is not required (Figure 4.28).

Error Handling

The *Error Handler.vi* intercepts error codes generated by the DLL in most low-level VIs and attaches string comments to the codes and then passes them on to a standard LabVIEW error cluster. Numeric error codes passed to the VI use the *Format Value* function. This is done because the DLL's error codes are nonsequential in nature and have many "missing" numbers. To avoid having to have cases for every error code number, both existing and missing, the incoming codes are more easily handled by string representations of their hex values. Some fatal errors are trapped by the VI

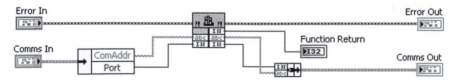

FIGURE 4.28 The *14443A Mode.vi*.

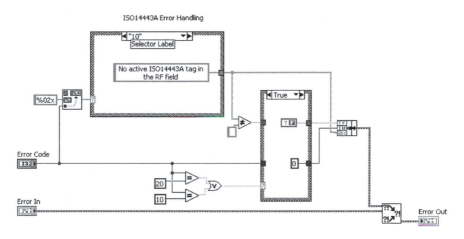

FIGURE 4.29 The *Error Handler.vi*.

in order to prevent the calling routines from crashing. These special case errors are related to an RFID card or tag not being in the RF field, which is a circumstance in the application that will routinely occur (Figure 4.29).

ISO15693

Although ISO15693 cards were not actually employed in this application, it was thought appropriate and useful to developers to describe VIs and methods that could be employed if users preferred this format over ISO14443A. The RR3036 DLL and the HF-TP-RW-USB1 reader both support 14443A and 15693 so no changes are required apart from the specific VIs related to the protocol used. It should be noted that 15693 cards can't be used if the reader is set to 14443A mode and vice versa. Any attempt to read a card in the wrong mode will cause the reader to report that no card is in the RF field.

The basic operations of 15693 are very similar in nature; however, some calls do vary a little from those of 14443A. Only those VIs that haven't previously been described for 14443A operation are related in the following sections.

ISO15693 Memory Organization

ISO15693 card memory is arranged in simple blocks, the size and number of blocks varying with card type and manufacturer. Two of the more common card types are the Texas Instruments Tag-It and the Philips I-Code cards. The TI device typically has 256 bytes arranged in 64 blocks and the Philips device has 112 bytes in 28 blocks, both devices possessing block sizes of 4 bytes. Due to the limited block sizes available in most of these early cards, a number of blocks may have to be used to maintain the same 16-character width used in this application. Refer to the "Memory Location.vi" in Figure 4.14 for further details on how this block linkage could be accomplished. If less precision is required, fewer blocks per variable can easily be accommodated.

6612 D3CD 0A80 **07**E0 is a typical "DSFIDandUID" number read from an ISO15693 tag (see the *Inventory.vi*). The number, as the name implies, consists of two discrete components, the DSFID and the UID. As previously described, the UID is a number unique to the particular RFID tag or card and is used to discriminate between individual tags. The data storage format identifier (DSFID) is consistent across all similar card versions and models. It contains two key pieces of data, the "Block Size" and "Block Num." These data define the available memory structure within the card. In the "DFSIDandUID" previously given, the **07** (07H) portion of the number defines the card's manufacturer, in this case "Texas Instruments." By referencing the following table we can also discern the card's "Block Size" is 4, and its "Block Num" is 64. As different manufacturer's cards' memory capacities vary, users should consult the relevant documentation carefully before deployment of particular card types and further application development.

	Manufacturer	Block Size	Block Num	Type	Other Code
00H	STMicroelectronics	4	16	B	
04H	Philips Semiconductor	4	28	B	E0H
05H	Infineon	4	58	B	60H
05H	Infineon	4	250	B	E0H
07H	Texas Instruments	4	64	A	E0H
08H	Fujitsu	8	250	A/B	
16H	EM Microelectronic	4	14	A	

BASIC RFID OPERATIONS

Process for Writing to an ISO15693 RFID Card

Writing ASCII text to the 15693 cards (Figure 4.30) involves the following process and order of execution:

1. Open com port (Auto Open Com Port.vi).

2. Get reader information, optional (Reader Info.vi).

3. Switch the reader to ISO15693 mode (Change to 15693.vi).

4. Switch on the RF field (Open RF.vi).

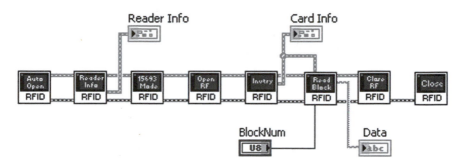

FIGURE 4.30 Generalized process for writing a string to a single ISO15693 tag block.

5. Get card information (Inventory.vi).

6. Write the string to the specified card block number (Write Block.vi).

7. Loop to #6 (above) as required.

8. Switch off the RF field (Close RF.vi).

9. Close com port (Close Com Port.vi).

Process for Reading from an ISO15693 RFID Card

Reading ASCII text strings from the ISO15693 cards (Figure 4.31) involves the following process and order of execution:

1. Open com port (Auto Open Com Port.vi).

2. Get reader Information, optional (Reader Info.vi).

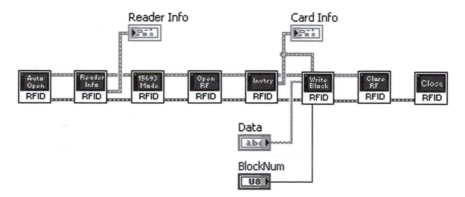

FIGURE 4.31 Generalized process for reading a string from a single ISO15693 tag block.

3. Switch the reader to ISO15693 mode (Change to 15693.vi).

4. Switch on the RF field (Open RF.vi).

5. Get card information (Inventory.vi).

6. Read the string from the specified block number (Read Block.vi).

7. Loop to #6 (above) as required.

8. Switch off the RF field (Close RF.vi).

9. Close com port (Close Com Port.vi).

Application High-Level VIs
Write Floating Point Numbers (and Strings)

The *Write Float.vi* (Figure 4.32) takes the floating point number passed to it by the "Data" control and converts it into a text string of 16-character length. The *Memory Locations.vi* takes the user's "Memory Location" passed to it and computes an array of block numbers that will be required to store each component of the 16-character text string. The text strings are written to the card using the "Write Block.vi." Immediately a block is written the block is read by the "Read Block.vi," and a comparison is made of the text intended to be stored with that recovered. A check is performed on each block, and if an error occurs, the "Write Error" Boolean will be made "True." If some truncation or rounding of the number occurs, the "Truncation/Rounding" Boolean will be "True" (note that some rounding of very high precision numbers may be expected).

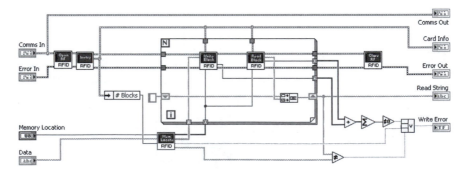

FIGURE 4.32   The *Write Float.vi.*

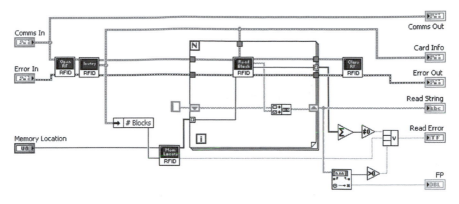

FIGURE 4.33 The *Read Float.vi*.

The *Write Strings.vi* is identical to the *Write Float.vi* except for the conversion to/from floating point data before the strings are written and output.

Read Floating Point Numbers (and Strings)

The *Read Float.vi* (Figure 4.33) takes the user "Memory Location" passed to it and hands it to the *Memory Location.vi* to compute an array of Block Numbers that will be used to read the various components of the athlete variable concerned. As each block is read a shift register concatenates the text strings until the variable is fully reconstructed. The complete string is then converted into its floating point equivalent. As with the *Write Float. vi* the "Comms In" and "Comms Out" are directly linked to prevent fatal errors if the card should not be present during the read operation.

The *Read Strings.vi* is identical to the *Read Float.vi* except for the final conversion of the output string.

15693 Master

The 15693 *Master.vi* enables a methodology for simplifying system development by enabling all functions to be called by placing a single VI on the user's wiring diagram. Like the *14443A Master.vi* it provides all available funtionality controlled by one enum. Figure 4.34 shows the *15693 Master.vi* in the "Get Card Info" case. This VI must be called once in the "Auto Open Com Port" state prior to performing subsequent functions. The communications port number and reader command address are remembered by the uninitialized shift register on the while loop. As with the *14443A Master.vi*, this VI uses the "Input Cluster" and "Output

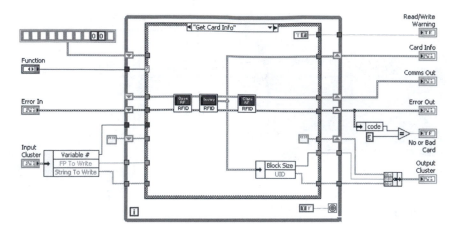

FIGURE 4.34   The *15693 Master.vi* shown in the "Get Card Info" case.

Cluster" for different purposes contingent upon the function being called. The functions that are available are as follows:

Detect reader and open communication port.

Get reader information.

Get RFID card info (e.g., UID, card type, manufacturer).

Read a stored 16-character ASCII string.

Write a 16-character ASCII string to card.

Read a stored floating point number.

Write a floating point number to card.

Read a section of card.

Read the entire card content.

Write a single block to the card.

Read a single block from the card.

Erase a card.

Beep the reader buzzer.

Flash the reader LED.

Close com port.

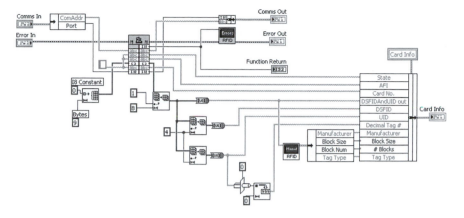

FIGURE 4.35   The *Inventory.vi*.

SubVIs

Inventory

The *Inventory.vi* shown in Figure 4.35 provides a mechanism for determining a variety of information about the currently active RFID card. In particular the VI returns the "DSFIDandUID" that is critical to all subsequent read and write operations. Importantly, this VI calls the *Manufacturer.vi*, which provides the card's "Block Number" and "Block Size." These data are required by the *Memory Locations.vi* to compute appropriate memory locations for the athlete variables used in the application. The *Inventory.vi* performs a similar role in 15693 as do the 3x 14443A commands: *Request.vi*, *Select.vi*, and *Anticol.vi*.

Memory Locations.vi

The *Memory Locations.vi* (Figure 4.36) is used to compute the requisite block numbers required to create the 16-character variables for the application. Depending upon the "Memory Location" and "Block Size" values passed to the VI, the Block #'s array will be filled with the needed sequential block numbers used to contain the components of each variable. This array data will be subsequently needed to perform the various read and write operations. If the card block size (as determined by the *Inventory.vi*) is 8, then 2 consecutive blocks are needed per variable; if = 4, then 4 will be required. In simple terms using this methodology, the Texas Instrument 64 block cards with a block size of 4 provide for 16×16 text character variables, whereas the Philips 28 block card provides for only 7. If the block

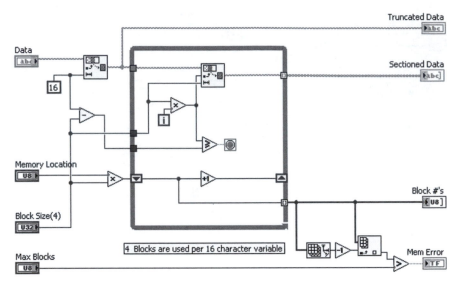

FIGURE 4.36   The *Memory Locations.vi.*

numbers generated in the "Block #'s" array exceed the card's capacity then "Mem Error" Boolean will be set to True.

Manufacturer

Figure 4.37 shows that the *Manufacturer.vi* extracts data from the DSFIDandUID string returned by the call to the Inventory function of the DLL. This VI provides the card's "Block Number" and "Block Size," data required by the *Memory Locations.vi* to compute appropriate memory locations for variables used in the application. It also extracts the manufacturer's code and provides string text of the firm's name.

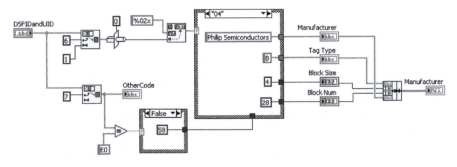

FIGURE 4.37   The *Manufacturer.vi.*

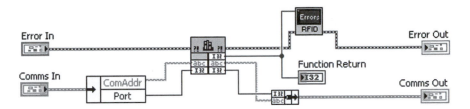

FIGURE 4.38   The *Change to 15693.vi.*

It should be noted that the "Comms Out" indicator is wired directly to the "Comms In" control rather than being connected to the *Close RF.vi* "Comms Out" connection (on the far right of the VI). The reason for this is that if during the write process the RFID card is either not present or is removed during the write an error is generated in the DLL. This error sets the communications port number and command address to their default values ("") causing subsequent attempts to write to fail. By wiring the "Comms Out" to the "Comms In" this otherwise fatal error is trapped and the card can be freely removed and replaced at will without issue.

Low-Level VIs
Switch to ISO15693 Mode
The *Change to 15693.vi* (Figure 4.38) switches the reader from any previously selected mode to ISO15693.

Open RF
The *Open RF.vi* (Figure 4.39) switches on the reader's RF field. Unless the field is operative no data can be retrieved from the cards.

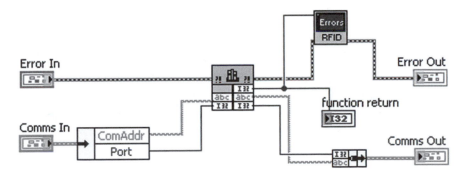

FIGURE 4.39   The *Open RF.vi.*

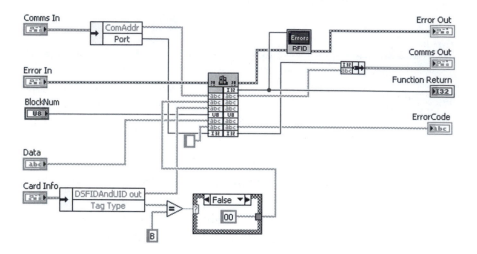

FIGURE 4.40   The *Write Block.vi.*

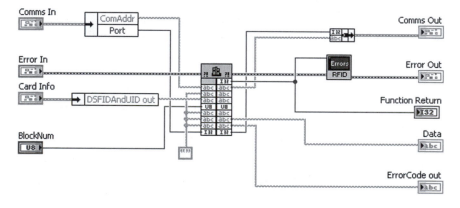

FIGURE 4.41   The *Read Block.vi.*

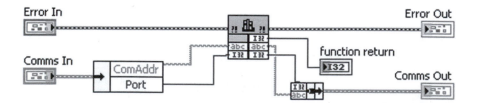

FIGURE 4.42   The *Close RF.vi.*

Write Block

The *Write Block.vi* shown in Figure 4.40 illustrates writing a 4 byte string to the block number passed to the VI by the "Block Num" control.

Read Block

The *Read.Block.vi* (Figure 4.41) reads a 4 byte string from the block number passed to the VI.

Close RF

The *Close RF.vi* (Figure 4.42) switches off the reader's RF field.

CHAPTER 5

Pachube

Sensing the World through the Internet

Anne M. Brumfield

CONTENTS

INTRODUCTION

The Internet of Things (IoT) refers to the connection of real-world objects to the Internet. Pachube (www.pachube.com) provides a vehicle for growing the Internet of Things through its online platform that delivers global visualization of sensor data streams. This novel architecture promotes data sharing, monitoring, mining, and scientific collaboration at little or no cost to its users. The functionality of Pachube is provided through its application program interface (API) that can be used from myriad

programming and scripting languages such as Java, Ruby, Pyth̶
PHP, Visual Basic, and LabVIEW. This chapter will introduce the ̶
to the Pachube web service and demonstrate how data feeds can be c̶
figured, updated, and distributed using LabVIEW and the current A̶
version.

BACKGROUND

The Pachube API, based on the RESTful web service model, provides methods for uploading and downloading data to and from the server using Hypertext Transfer Protocol (HTTP) requests. Thus, the data can be served, accessed, and analyzed by independent applications customized to meet the user's needs. The reader is encouraged to visit the Pachube website to peruse the libraries, code examples, and tutorials that will not be detailed in this chapter. Content will instead focus on parts of the API that are necessary for serving and retrieving data, specifically from within the LabVIEW environment.

The Pachube Environment

The Pachube data hierarchy consists of an environment or feed composed of one or more data streams. The data stream, which represents data from a particular sensor or device, consists of data points or pairs of time-stamped values. Data streams can be referred to by their ID or by a tag or user-defined name.

Upon entering the website, one is greeted with a world map view of user feeds (Figure 5.1). These can be browsed by location and type for additional information and, with registration, can be accessed to retrieve real-time data. The registered user is eligible to create and access up to five feeds free of charge at any given time (feeds may be deleted and replaced), although there are limitations on API transactions (five requests per minute) and a cap on history data at the one-month limit. Additional usage requires a modest subscription fee but provides a year of historical data storage, up to 40 data streams, and 40 API requests per minute. By registering as a basic user, experimentation with the API and the environment can be done at no cost while configuring sensors and assessing the utility of porting sensor feeds or connecting to other users' Pachube outputs. This feature alone will make this website a temptation that is difficult to resist for most readers.

Once verified with a user name and password, a master API key is provided that allows users to access their data streams (within the confines

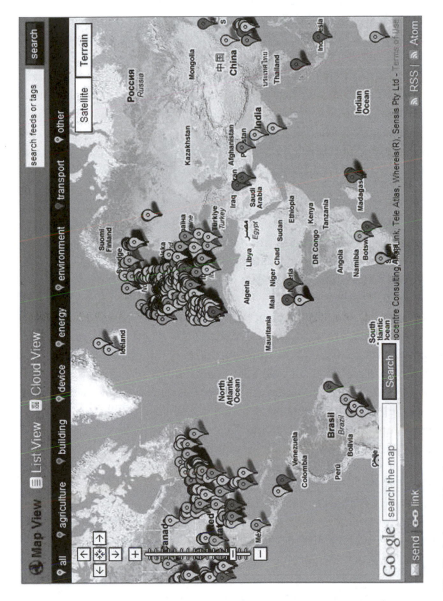

FIGURE 5.1 The Pachube worldwide community.

of their account type and permissions). Without access to the API key, feed updates are available every 15 minutes. Upon login, users can customize their webpage and define their feeds, environment, and location as shown in Figure 5.2. A feed has been created known as "Physiology" and has been given the ID 21881 by the Pachube site. Two data streams, with the tags SpO2 (id=0) and Heart Rate (id=1), make up the content of this feed.

Sensor data from feeds may be viewed on the Pachube website; however, many tools are available that allow users to embed customizable visualizations of their data in their own web pages. The site's apps section includes configurable, zoomable graphs, dials, and dashboard gadgets (Figure 5.3). There are email, Twitter, and short message service (SMS) alerts that can be configured using threshold triggers as well as applications ready to use for iPhone, Android, and Blackberry devices.

Communicating with the Pachube Environment

Extended Environments Markup Language (EEML) provides a protocol for exchanging and sharing sensor data between environments in a structured manner such as using data-stream tagging (www.eeml.org). The reader may be familiar with the data modeling language known as eXtensible Markup Language (XML), which is used to describe EEML. The API supports, from simplest to complex, Comma-Separated Values (CSV), JavaScript Object Notation (JSON), and EEML. There are advantages to using one over another depending on the user's application. JSON and EEML contain metadata, or data about data content, a feature that is useful when integrating Pachube with web applications. Embedded devices such as the Arduino microcontroller make use of the simpler CSV format for communication with Pachube. The various formats will be demonstrated in the LabVIEW examples that follow.

CHALLENGE

The challenge of this project was to provide a demonstration of the interaction between LabVIEW graphical programming and the Pachube web server that provides real-time visualization of sensor data.

SOLUTION

Communicating from the LabVIEW Environment

The .NET Framework provides a simple solution via the WebClient class for sending and receiving data across network resources. This language-independent, object-oriented approach allows users to manage requests

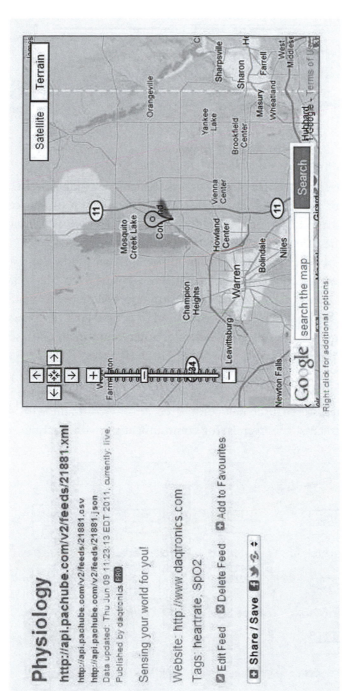

FIGURE 5.2   Sample user-defined webpage.

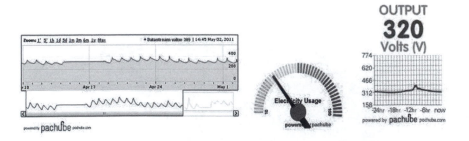

FIGURE 5.3   Pachube embeddable graphics.

and responses using strings, byte arrays, streams, or files. The LabVIEW environment provides functions that allow the creation of and communication with .NET objects, provided that the latest .NET Framework is installed. The palette of .NET functions is shown in Figure 5.4.

The components that will be used in the VIs that follow consist of the Constructor, which initializes a new instance of the WebClient class, the Property Node, which reads or writes properties such as the header, and the Invoke Node, which invokes a method or action such as uploading or downloading data. Recall that the header is necessary in all of these VIs as it provides the API key that is necessary for feed access.

Data Format Structure

Once the registered user has configured an environment with sensors and data streams, the exercise of communicating with Pachube ensues.

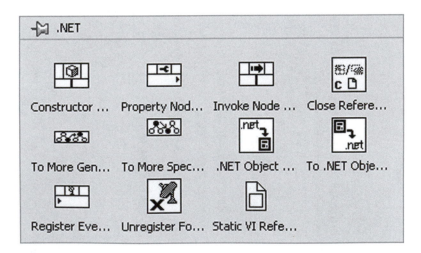

FIGURE 5.4   LabVIEW's .NET palette.

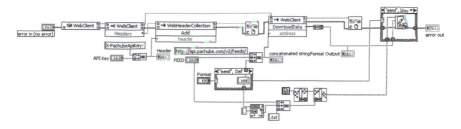

FIGURE 5.5   Retrieve Data Format Structure.vi.

As discussed already, there are three data formats to accomplish this, and the appropriate choice depends on the application. The *Retrieve Data Format Structure.vi* provides a means for visualizing the content and syntax of each particular data format (Figure 5.5).

In this instance, the DownloadData method of the WebClient class is invoked, along with an address of the form http://api.pachube.com/v2/feeds/<insert feed id>.xxx, where xxx can be xml, json, or csv (e.g., http://api.pachube.com/v2.feeds/21881.xml). The resultant output is saved to <feed id>xxx.txt (e.g., 21881xml.txt) in the current directory. Since the versions of json and xml change from time to time, this can be a useful means for retrieving the current structure for a new feed. This structure can subsequently be used to upload data and in the case of json and xml, conveying essential metadata, such as minimum and maximum value limits. An example of uploading string data in json format is shown in Figure 5.6.

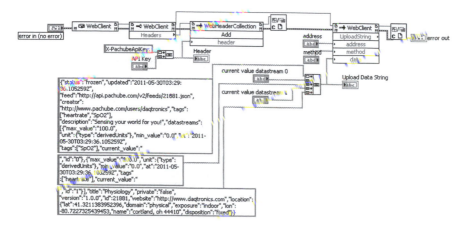

FIGURE 5.6   Test JSON Upload.vi.

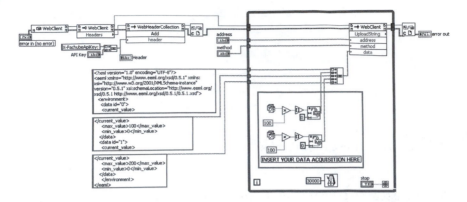

FIGURE 5.7 Uploading simulated data.

Serving Data to Pachube

Feeds may be configured for manual or automatic updates, with data being pushed or pulled, respectively. The example in Figure 5.7 demonstrates a manual feed configuration. In this instance, it is a simple update of randomly generated numbers within the loop. This provides a quick way to assess a new Pachube environment without having to configure sensors. Waveform generation functions could also be placed inside the loop to simulate predictable data and to allow communications to be evaluated. This portion of the loop can be replaced with data acquisition code to upload sensor outputs easily enough.

Other Useful Tools

As readers begin to experiment with their own feeds, it may be useful and meaningful to look at the history or archive data of published feeds, those owned, as well as those belonging to others in the community, particularly if there are other users providing data from the same sensors or similar applications. The *Get All Feeds.vi* (Figure 5.8) will provide a listing

FIGURE 5.8 API call to get all feeds viewable.

of all feeds viewable by the authenticated account. Again, the user may select the format of the output as csv, xml, or json. Feeds may be searched using descriptors such as tag=humidity or q=arduino or status=live. The API provides additional documentation on this feature. Such a search will provide any feeds matching the criteria, along with their associated data streams, latest time stamp, and value.

The ID of any feed can be used to plot the history, the last 24 hours of data in 15-minute increments, or the archive, which is all of the data since the start of the feed in 15-minute increments. The front panel of the *Retrieve Archive_History Data.vi* is shown in Figure 5.9. The history or archive plot can be chosen from the menu.

Similarly, a Portable Networks Graphics (PNG) image file of the history data for a given feed may be retrieved by invoking the DownloadFile method of the WebClient class. As usual, the address is supplied; however, in this case it can be amended to set the plot color, size, graph and legend

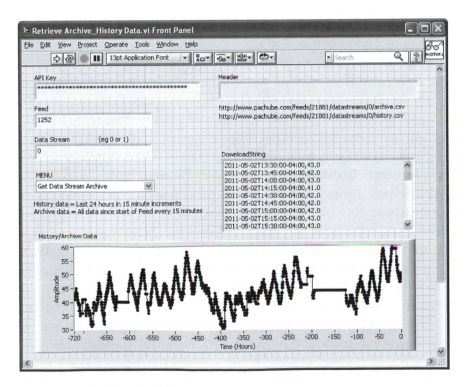

FIGURE 5.9   A plot of archive data.

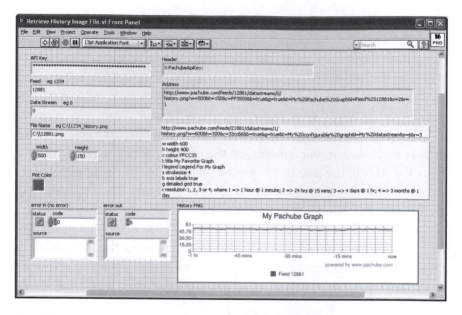

FIGURE 5.10   Retrieving a PNG image file from Pachube.

title, and a variety of other features including the amount of data retrieved. The front panel and the block diagram for the *Retrieve History Image File. vi* are shown in Figure 5.10. The PNG image is opened and displayed on the panel, and the various features that can be customized are indicated.

Another useful feature is the ability to create, read, update, and delete triggers. Triggers are configured to send HTTP POST requests to a designated Uniform Resource Locator (URL) of the user's choice when threshold criteria are exceeded or met. This is particularly useful for monitoring an environment to receive an email, Tweet, or alarm when specified conditions are encountered. For trial purposes, the POSTBIN web service can be used in lieu of the user's own URL to test the triggering functionality. Go to http://www.postbin.org and click "Make a Postbin" to be redirected to a temporary URL that will serve as a log for POST requests that result from the execution of a trigger. Manually configure a trigger by clicking on the "embed, history, triggers, etc" under the data stream plots on the Pachube web interface and enter this URL along with the threshold criteria. Select the "Debug" option for each configured trigger to POST to the specified URL, and verify the functionality of each by returning to the URL site and refreshing. Note that each trigger is given an ID.

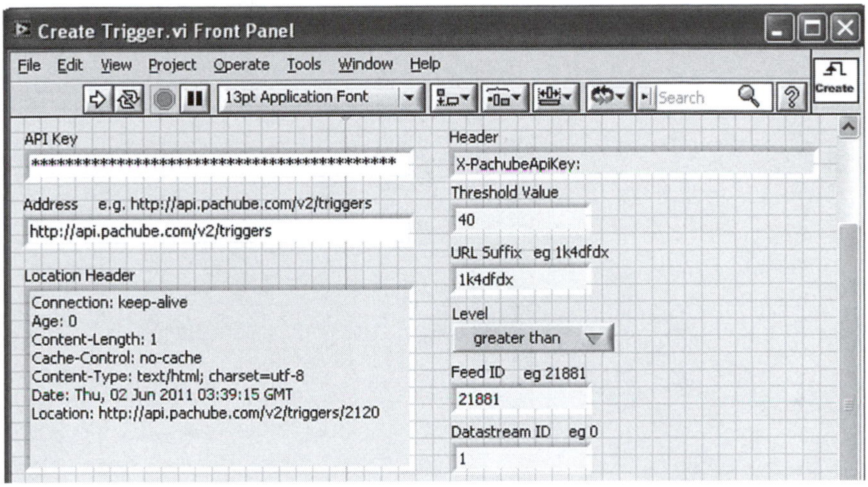

FIGURE 5.11   Create Trigger.vi.

Obviously, setting up a manual trigger is inferior to using the API to handle triggers programmatically. Examples for handling triggers from within LabVIEW that make use of the API should shed some additional light on the advantages of using the .NET approach. The *Get All Triggers. vi* provides a list of all configured triggers and can be used to verify any triggers that have been configured manually in the previous exercise. The *Create Trigger.vi* is shown in Figure 5.11 and provides the trigger ID in the ResponseHeaders string output. Note that a URL must be designated in the configuration, either the user's own or one of the POSTBIN nature already described. One should be able to confirm trigger IDs found in the Responseheaders output with those found posted to the URL.

The *Delete Trigger.vi* is shown in Figure 5.12. This is a permanent method and can be verified by refreshing the Pachube feeds web page. With these three VIs, *Get All Triggers.vi*, *Create Trigger.vi*, and *Delete Trigger.vi*, the user should be able to experiment with programmatic creation, deletion, and verification of configured triggers.

CONCLUSION

The VIs in this chapter touch upon some of the methods for interacting with Pachube from the LabVIEW environment. The latest API version is available on the Pachube site and includes many more features that are offered to the user such as programmatically creating and deleting users,

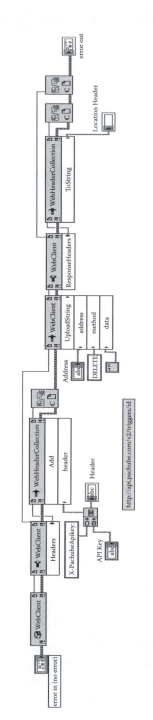

FIGURE 5.12   Delete Trigger.vi.

feeds, data streams, and keys. Hopefully this chapter has served as an introduction of the application of .NET for providing communication between LabVIEW and Pachube.

Happy sensing!

RESOURCES

National Instruments 2010 or higher

.NET Framework 3.5 SP1

Pachube Basic User Account (www.pachube.com)

REFERENCES

Microsoft. WebClient Class. http://msdn.microsoft.com/en-us/library/system.net.webclient(v=vs.80).aspx. Retrieved May 6, 2011.

Pachube, Patching the Planet: Interview with Usman Haque. Ugotrade, by Tish Shute. http://www.ugotrade.com/2009/01/28/pachube-patching-the-planet-interview-with-usman-haque/.

Power System Applications in LabVIEW

Nesimi Ertugrul

CONTENTS

INTRODUCTION

Power systems cover a wide range of diverse applications such as rotating electrical machines, renewable energy systems, power electronics, and distribution systems, and they display distinct electrical behaviors. In addition, such systems are usually interrelated and imply highly nonlinear operating characteristics. Although the operators and designers have a good degree of knowledge about the steady-state behavior and the frequency and magnitude of measurable parameters (e.g., voltage, current, speed, torque) of the power systems, various unpredictable signals may also be present specifically during initial states of electrical circuits. For example, inrush currents in transformer and starting torque and current characteristics of rotating machines heavily depend on the initial switching instants as well as the load level, which cannot be predicted. Moreover, monitored signals may have unpredictable characteristics specifically in power quality measurements in a large power grid, which occur randomly. Furthermore, the frequency bandwidth of measured signals in power systems usually covers a very wide range from direct current (DC) to 10s of kHz. Therefore, it can be concluded that the measurement requirements in power systems are unique from the signal characteristic and the analysis viewpoint and require special attention.

LabVIEW has opened a new paradigm and contributed to the progress and growth of power system related studies, and it is now the most valuable member of the mainstream enabling tools in power systems. LabVIEW can demonstrate and represent various electric diagrams and can considerably enhance understanding through the adequate exploitation of resources available in full-graphics screens. In addition, it provides analysis tools to develop highly complex systems with accurate measurement and control capabilities.

The integration of basic concepts, visualization, and automation capabilities of LabVIEW opened a new horizon to power system applications. An object-oriented information model of the components of power systems in LabVIEW is now commonly used to give context and meaning to data. Moreover, the measurement capabilities of LabVIEW provide new avenues to understanding, improving integration and interoperability of various components of power systems, enabling accurate automation of setup, and even offering predicted maintenance tasks such as in power quality and electrical machine monitoring.

BACKGROUND

A large number of LabVIEW-based studies have been reported in the literature from basic electric circuit simulation activities to complex motor control systems. A limited number of studies have also been reported where LabVIEW is used to monitor the behavior of a power system and to determine its response involving very large electrical machines operating under normal as well as faulty conditions.

The studies reported in this chapter are a part of the mainstream LabVIEW-based research and consultancy activities at the University of Adelaide covering a wide range of power system related applications. This chapter aims to provide a guide to the users of power systems who are considering developing related systems as well as to share some of the experiences gained during the software and hardware development stages at the University of Adelaide. The applications covered here are unique since their operating conditions cover almost the entire operating range of a power system involving low-level and high-frequency measurements in electrical machines and in a power grid. Although only five major applications are included, they have common characteristics, can easily be extended to other applications, and have room for improvement and customization. These LabVIEW applications are:

- Power quality monitoring in power systems
- Condition monitoring of induction machines
    - A low-cost continuous condition monitoring unit with CompactRIO
    - PC-based online monitoring system
- Real-time solar array monitoring in remote areas
- Automated dynamometer test setup
- Electrical machines tests

In the following sections of this chapter, enhanced representations of the developed applications along with the relevant front panels will be provided together with the principles of the implementations and measurement methods. The primary objective is to help readers understand the characteristics of a given power system and allow a better interpretation of programs and associated data and results while improving the productivity and offering common approaches for future developments.

CHALLENGE

From the implementation viewpoint, the distinguishing features of any LabVIEW-based measurement and control mechanism in power systems are the necessity of electrical isolation and the requirements for the wide frequency bandwidth measurements. These impose significant restrictions on the accuracy of measurements that have to be integrated into sensors, signal conditioning devices, and also custom VIs.

The electrical isolation is the most critical requirement for the safety of equipment and operators in power systems. However, when considering a sensor, its characteristics need to be assessed based on three primary criteria: (1) frequency bandwidth; (2) magnitude of the output signal; and (3) electrical isolation.

For example, although the fundamental frequency of an alternating current (AC) supply is very low (50 Hz, 60 Hz), condition monitoring of electrical machines and power quality monitoring require monitoring signals up to 10 kHz. In addition, a conventional voltage transformer may be sufficient to provide a good electrical isolation, but it also filters higher-frequency signals, which are the primary interests in such systems.

In principle, measurements in power systems involve voltage, current, flux, vibration, torque (or force), and speed (or position) measurements, and a form of signal conditioning device is required to obtain voltage output that is proportional to these parameters. In addition, the signal conditioning device usually provides a degree of amplification to provide a voltage output that varies within the input limits of the data acquisition system.

Furthermore, it is important to test and calibrate any sensor and associated signal conditioning circuit before connecting to a data acquisition system. During the calibration stage, a signal with a known frequency and amplitude is applied to a sensor/transducer using a signal generator or a known load, and its output is measured by some other means (e.g., using an oscilloscope or a true root mean square [RMS] voltmeter). The gains of the amplifiers and the gain of voltage divider, such as in isolation amplifiers used to measure high voltages, should also be considered in the calibration tests. It is desirable to make the calibration test for each measurement channel. Note that LabVIEW-based programming can allow the user to integrate the calibration results into the main measurement.

Table 6.1 summarizes the main quantities that can be measured in power systems as well as their associated characteristics.

TABLE 6.1 Quantities That Can Be Measured in Power Systems and Basic Sensor Specifications

Quantity	Device	Function and Remarks	Typical Bandwidth	Typical Input Range	Typical Output Gain
Voltage	Differential amplifier	For high-voltage measurements, but not real electrical isolation	DC to MHz	± 600 V	100 V = 1 V
	Isolation amplifier	Safe high-voltage measurements	DC to 50 kHz	± 1000 V	100 V = 1 V
Current	Hall-Effect current transducer	Detects magnetic field produced by current, available in clamp type or through-hole	50 kHz	± 10 A	10 A = 1 V
	Rogowski coil	With a flexible cord suitable for measuring large diameter cables and rotating shafts	few kHZ	± 1000 A	100 A = 1 V
Flux	Search coil	Measures flux produced by currents such as the current flowing in the stator and rotor end-windings	10 kHz	± 1 V	Proportional to the number of turns
Vibration	Piezoelectric accelerometer	Acceleration measurements	20 kHz	± 2 g	1 g = 1 V
Speed	Tachometer	Measures shaft speed directly	multiples of kHz	0 rpm >2000 rpm	
Torque	In-line torque transducer	Measures shaft torque directly	multiples of kHZ	Up to 50 Nm	0.1 V/Nm
	Force gauge	Measures average shaft torque indirectly	below kHZ	Up to 100 Nm	0.1 V/Nm

SOLUTIONS

In the following sections, five different LabVIEW-based power systems applications are described. In each application, the structure of the approach is explained, followed by the associated front panels. Justification for using LabVIEW in each application is also briefly given.

Power Quality Monitoring in Power Systems

Since it is known that the AC power grid has three phases, an ideal three-phase AC supply consists of three phase voltages that are 120 degrees out of phase with identical magnitudes. Of course the primary requirement is that these voltages should have sinusoidal waveform characteristics at constant frequency and should be available continuously. Any diversion from these requirements to a level that has an adverse effect on the electricity consumers is considered to be poor-quality power or *polluted*.

As a general statement, power quality can be referred to as the degree to which voltages and currents in a power system represent sinusoidal waveforms. *Clean* power refers to voltage and current waveforms that represent pure sine waves and are free from any distortion, whereas *polluted* power refers to voltage and current waveforms that are distorted and cannot be represented by pure sine waves. Voltage dips (sags), swells, interruptions, switching transients, harmonics, notches, or flickers are examples of the most frequent disturbances in power system networks (Figure 6.1).

The motivation and the objectives of this LabVIEW application are directly related to the power quality problems (disturbances) faced in the industrial world as electric power systems have become more polluted than ever before. This issue is primarily related to ever-increasing sources of disturbances that occur in interconnected power grids, which accommodate a large number of power sources, transmission systems, transformers, and interconnected loads. In addition, such systems are

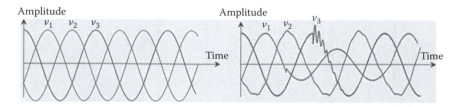

FIGURE 6.1 Ideal, "clean," three-phase AC waveforms and hypothetical "polluted" AC waveforms with voltage sag (phase 1), with harmonics (phase 2), and with transients (phase 3).

exposed to environmental disturbances like lightning strikes and storms. Furthermore, nonlinear power electronic loads, such as converter-driven equipment (from consumer electronics and computers up to adjustable speed drives), have become increasingly common in power systems. Although these pieces of converter-driven equipment are manufactured according to the associated standards, the wide use of such devices pollutes the power systems. If these unwanted variations in the voltage and current signals are not mitigated properly, they can lead to failures or malfunctions of the many sensitive loads connected to the same systems, which can threaten the entire system security and may be very costly for the end users. Although it is not easy to estimate accurately the exact cost of the pollution in the quality of the power, poor power quality just in the United States causes about US$13.3 billion in damage per year.

It is important to emphasize here that the structure of the power systems from the generation to the end users has become more private and belongs to different utilities. Unlike earlier structures, however, the current power grids have a number of stakeholders, and it is no longer clear who is responsible for the reliability and the quality of the AC supply provided to customers. Therefore, for the sake of improvement it is highly critical to identify the source of the disturbances and the responsible stakeholders. Numerous studies examine ways of reducing the impact of the power quality aspects mainly under two groups: (1) custom power solutions (mainly hardware oriented); and (2) power quality monitoring solutions.

However, due to the large number of different types of disturbances and the ever increasing number of power quality monitors installed in power systems, manual analysis of captured disturbance is no longer a practical option. Ideally, it is desirable to have automatic analysis tools that are integrated with the monitoring systems and can be applied to large disturbance databases to cluster them automatically. Developing an accurate automatic power quality monitoring system, however, is highly complex and primarily involves three main sections: (1) triggering; (2) feature extraction; and (3) classification (Figure 6.2). The LabVIEW solution explained here aims to address the triggering part of the power quality monitoring solutions that will be described as follows.

Since the power quality disturbances usually affect the voltage or current signals in different ways, different disturbances may require different triggering mechanisms. As a result, the existing triggering techniques may not capture every type of disturbance. Therefore, the LabVIEW solution

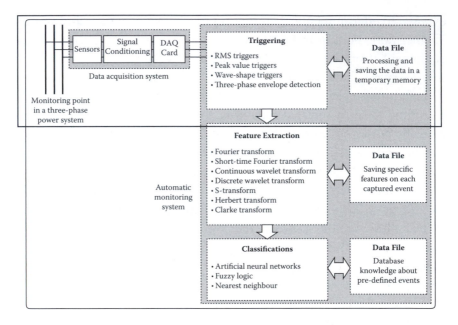

FIGURE 6.2   A block diagram highlights the three main steps of automatic monitoring process.

in this section offers an appropriate triggering mechanism that can be implemented in a custom-developed monitoring program to capture a wider range of disturbances.

Software Section

In the software section, a custom monitoring program for tracking real-time distortions in voltage and current signals was developed based on the point-by-point three-phase signal envelope in LabVIEW. The hierarchal structure of the main subVIs in the program are shown in Figure 6.3. As shown in the figure, the main VI communicates with five subVIs. In the "Configuration VI," the specifications of the hardware are configured and defined. The "File setup VI" specifies the type of each data from the input/output (I/O) file and adjusts the gain of the acquired data. The "Read Data VI" allows the system to start acquiring the data from the data acquisition (DAQ) card. In the "Data Processing VI," the data are diagnosed for any out-of-tolerance condition. Finally, in the "Save Data VI," the information about any distorted signal is saved in a specific memory.

Each of the subVIs in this application consists of a block diagram, which represents the structure of the program, and a custom front panel interface.

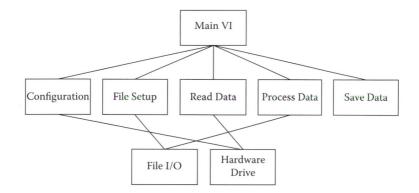

FIGURE 6.3   Block diagram of the LabVIEW monitoring system.

The main VI block diagram and its front panel are shown in Figures 6.4 and 6.5, respectively, where the voltage or current signals are continuously monitored for any out-of-tolerance in signals. If a predefined limit of the three-phase envelope is exceeded, the program saves the signal in a temporary memory for further analysis, which includes the polluted signal.

Resources

The LabVIEW-based power quality monitoring system has both hardware and software components for capturing real-time power quality signals. The general schematic diagram of the developed monitoring system is shown in Figure 6.6. The PCC in Figure 6.6 denotes the Point of Common Connection, which is the point of connection of the monitoring system.

The initial testing of the developed module can be done in a laboratory environment (Figure 6.7) for capturing a number of experimentally simulated power quality signals. To generate various disturbances, different power system configurations can also be arranged. In these configurations, a 415 V three-phase power source can be used to connect to different loads, such as resistors, induction motors, capacitors, and transformers, to introduce a degree of power quality events at PCC and to capture the effect of switching and resulting power disturbances.

The hardware section used in the application is composed of a DAQ card, which is hosted by a personal computer, and the voltage and current sensors. The specifications of the voltage and current sensors used are presented in Table 6.2.

Note that the sensor ratings given in this section are designed to perform tests directly on standard three-phase AC mains supply. If the tests

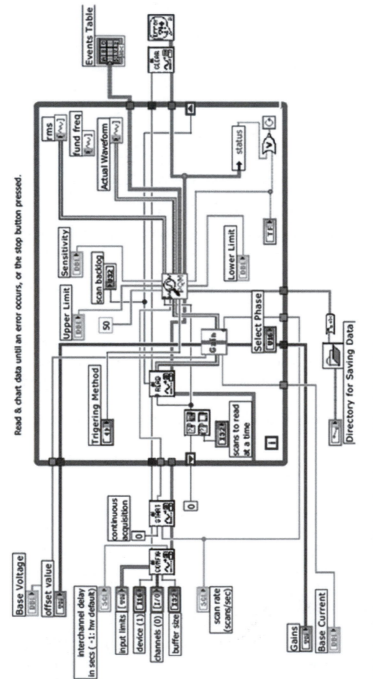

FIGURE 6.4 Block diagram of the main VI in the monitoring system software.

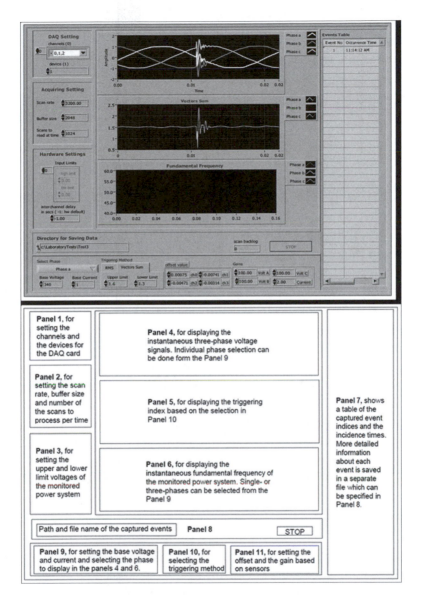

FIGURE 6.5   Front panel of the monitoring system (top), and an explanation of the function of each of the subpanels (bottom).

are going to be carried out at higher-voltage distribution systems, both standard current transformers and voltage transformers on the main switch boards may be used to be able to access the high voltage and the high current measurements. However, the frequency bandwidth of the transformers needs to be checked before performing any measurements.

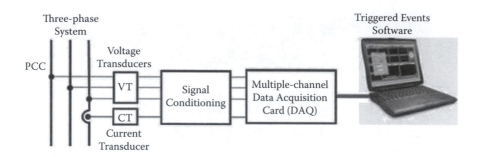

FIGURE 6.6   General block diagram of the monitoring system.

Condition Monitoring of Induction Machines

Induction motors are the workhorses of industrialized economies, and it is estimated that in developed countries they consume nearly half of all electric energy generated. There are about 10 million large (>80 kW) induction motors worldwide, which are generally very reliable and require minimum maintenance. However, duty cycle, installation, manufacturing factors, and environmental influences can deteriorate and reduce the efficiency of these motors as in the other rotating machines. In extreme cases, motors may have unexpected downtime, which can significantly increase the running cost or even be catastrophic in mission critical applications. In other cases, the efficiency of the motors may have been reduced due to the presence of various motor faults but the motors may still operate. This kind of failure may have a significant cost if not handled promptly and properly. For example, if the efficiency of an induction machine is reduced from 80 to 70%, an additional electricity cost of about 14% per year would be produced.

Conventional condition monitoring equipment for detecting faults in electrical machines is very expensive, and thus continuous condition monitoring techniques have been economical only for large, critical equipment such as power station generators. Two distinct LabVIEW-based induction machine monitoring systems have been developed and are explained in this section as an alternative to the commercial devices, which can offer high-accuracy measurement and analysis tools at reasonable cost.

A Low-Cost Continuous Condition Monitoring Unit with CompactRIO
Recently, due to the development of low-cost compact reconfigurable devices, it has become possible to develop a reconfigurable device to be used in industrial applications, including online monitoring of induction

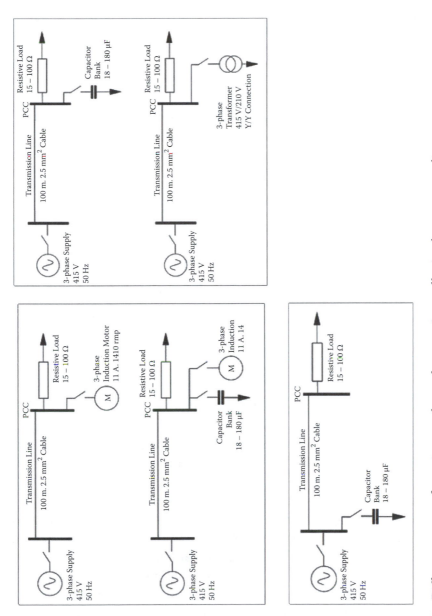

FIGURE 6.7 Different setup configurations can be used to experimentally simulate power quality events.

TABLE 6.2   DAQ Card, Voltage, and Current Sensors Specifications

DAQ Card Specifications	
National Instruments	
4 analog inputs	
12 bit analog-to-digital converter (ADC) per channel	
Up to 10 MS/s per channel sample rate	
Analog and digital triggering	
AC or direct current (DC) coupling	
8 input ranges from ±200 mV to ±42 V	
Voltage Sensor Specifications	
Bandwidth	DC to 15 MhZ (−3 dB)
Input attenuation ratio	Between 1/10 and 1/100
Maximum allowed differential voltage	±500 V (DC,AC$_{peak}$) or 350 V$_{rms}$ (1/100)
Maximum common mode input voltage	±500 V (DC,AC$_{peak}$) or 350 V$_{rms}$ (1/10, 1/100)
Current Sensor Specifications	
Bandwidth	Dc to 100 kHz
Current range	20A DC/30A AC
Accuracy	±1% 2 mA
Dielectric strength	3.7 kV, 50 Hz, 1 min
Output sensitivity	100 mV/A
Resolution	±1 mA
Load impedance	>100 kΩ

motors. The overall concept in this approach is to develop a low-cost continuous monitoring unit, called an indicator unit, based on low-cost sensors and a high-performance stand-alone processing system, CompactRIO. The major benefits of the continuous monitoring process are a reduction of the number of unexpected failures due to fault detection at an earlier stage and elimination of the need for periodic condition monitoring inspections. It can be emphasized here that for low- and medium-voltage motors the typical cost of a motor inspection is about US $500 per motor, and it is common to have these tests at six-month intervals, giving an annual cost of US $1000. Depending on the location, travel costs typically add about 50% to the testing cost, resulting in a rough annual inspection cost of about US$1500. This annual cost is of the order of the initial target cost for the indicator unit.

A conceptual diagram of the indicator box, possible sensor types, and their locations are illustrated in the left side of Figure 6.8. As indicated

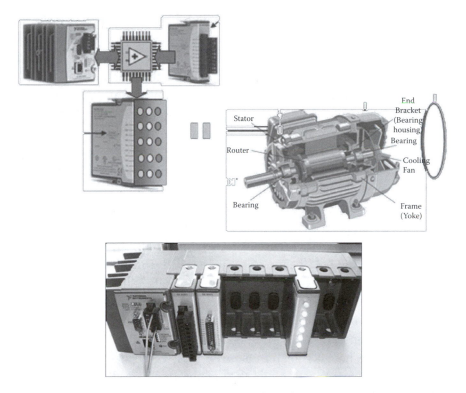

FIGURE 6.8 Conceptual CompactRIO based condition monitoring system diagram with sensors (top left), and CompactRIO with the developed (top right).

in the figure, the system continuously monitors the motor using a set of voltage, current, vibration, and flux sensors. If a fault is detected, the indicator unit alerts the nearby operator and shows the fault type and level using a series of lights on its front panel (right side of Figure 6.8). Note that if needed a more detailed fault description can be obtained from saved data.

Presentation functions, which are simple light-emitting diode (LED) indicators, and are added to the indicator to communicate the results with operators. The development basically has three major parts: (1) the CompactRIO indicator unit; (2) the CompactRIO field-programmable gate array (FPGA); and (3) the CompactRIO real-time controller software. To meet the application requirement, three basic modules are required for a CompactRIO base unit: (1) CompactRIO real-time embedded controller; (2) CompactRIO chassis; and (3) I/O modules.

The CompactRIO base unit comprises an FPGA and real-time controller. The CompactRIO real-time embedded controller (cRIO-9004) has the capacity to perform real-time stand-alone execution applications. To support these capabilities, the real-time controller provides 512 Mb of nonvolatile storage memory, 64 Mb random access memory (RAM), and a 200 MHz Pentium processor. For this research, the Ethernet port is connected to a PC as a reengineering and development station for uploading FPGA bit files and condition monitoring software. Therefore, based on these potential capabilities, the CompactRIO real-time controller is employed to perform real-time execution of the condition monitoring application including windowing, Fast Fourier Transform (FFT) analysis, averaging, peak detection, and fault analysis.

A custom-made CompactRIO LED-based indicator module, which is basically not part of the CompactRIO based unit, is also developed in this application. This module functions as a simple indicator using LEDs, as displayed in Figure 6.8 (the module on the right). This module uses Serial Peripheral Interface (SPI) protocol to communicate with the FPGA.

The data analysis software is performed in the real-time host controller cRIO-9004, which is basically divided into three main parts: (1) host to FPGA synchronization; (2) frequency analysis; and (3) fault analysis. A detailed flowchart is shown in Figure 6.9. As can be seen in the figure, the initialization connection and data calibration including engineering unit conversion of each sensor signal are performed first. The analog sensor signals are then captured, followed by an FFT analysis, frequency averaging, and fault frequency peak detection. Following the peak detection algorithm, a number of subroutines are implemented to identify specific faults. All the custom-written LabVIEW subVIs with brief explanations are given in Table 6.3.

Each VI is part of the fault analysis and diagnosis section in the CompactRIO real-time monitoring software. The CompactRIO condition monitoring system is used to identify broken rotor bar, eccentricity, stator shorted turns, and soft foot faults. A 2000 Hz sampling frequency is used, and 4096 samples for each channel was found sufficient in this system.

Figure 6.10 shows the LabVIEW front panel of the developed CompactRIO-based condition monitoring system. This front panel illustrates the single-phase motor current and corresponding frequency spectrum. In this study the programmable FPGA cRIO-9104 (Xilinx Virtex II XC2V3000-4FG676I, 3 million gate sizes, eight I/O module slots) chassis is used, including the signal processing function such as implementing low pass digital filtering routines. Furthermore, the FPGA chassis also reads sensor signals directly from the analog input module. The FPGA writes

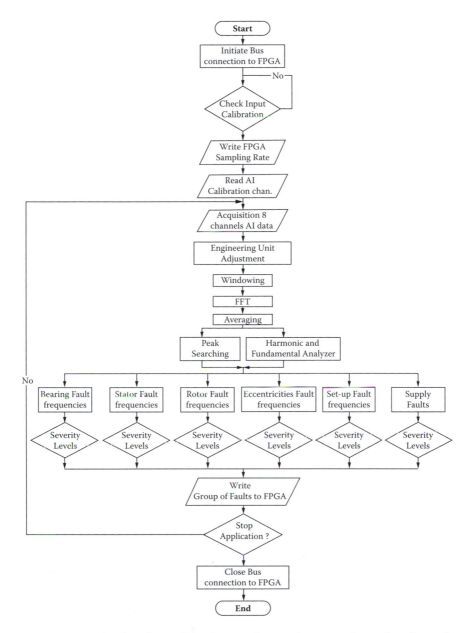

FIGURE 6.9   The flowchart of the data analysis software performed in the real-time host controller cRIO-9004.

TABLE 6.3   The Custom-Built VIs Implemented for the Machine Fault Frequency Analysis

Faults	Descriptions

Bearing Faults

To detect bearing fault frequencies

Bearing Faults

Bearing faults

Bearing related faults with stator current, voltage, and flux leakage

Stator Faults

Stator faults with vibration monitoring

Stator faults

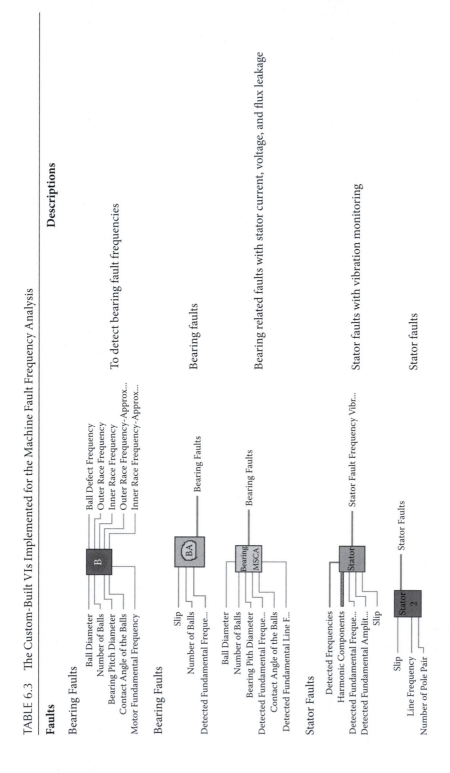

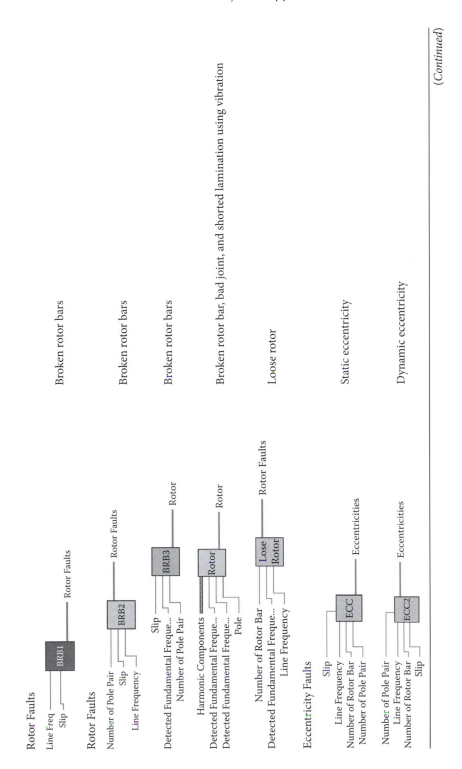

(Continued)

TABLE 6.3 The Custom-Built VIs Implemented for the Machine Fault Frequency Analysis (Continued)

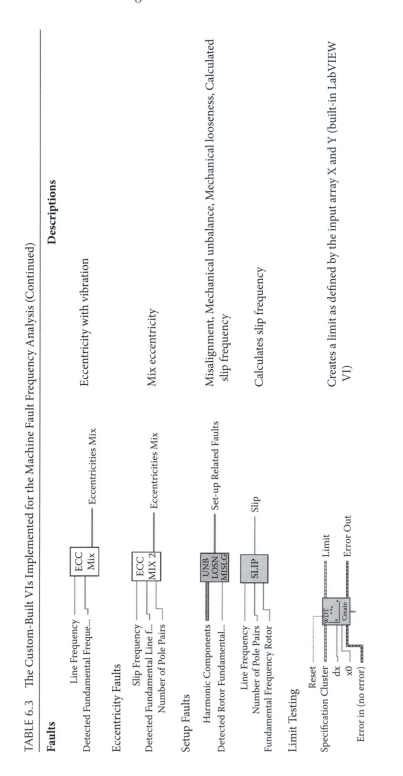

Faults	Descriptions
Line Frequency, Detected Fundamental Freque... → ECC Mix → Eccentricities Mix	Eccentricity with vibration
Eccentricity Faults	
Slip Frequency, Detected Fundamental Line f..., Number of Pole Pairs → ECC MIX 2 → Eccentricities Mix	Mix eccentricity
Setup Faults	
Harmonic Components, Detected Rotor Fundamental... → UNB LOSN MISLG → Set-up Related Faults	Misalignment, Mechanical unbalance, Mechanical looseness, Calculated slip frequency
Line Frequency, Number of Pole Pairs, Fundamental Frequency Rotor → SLIP → Slip	Calculates slip frequency
Limit Testing	
Reset, Specification Cluster, dx, x0, Error in (no error) → WDT Create → Limit, Error Out	Creates a limit as defined by the input array X and Y (built-in LabVIEW VI)

Limit Testing

Performs limit testing on cluster input data: compares **signal** with **upper limit** and **lower limit** (built-in LabVIEW VI)

Peak Searching and Fundamental Detection

Performs peak search on spectrum or spectra, single/multiple peaks

Detects fundamental and harmonics frequency including total harmonic distortion (THD)

Limit Testing

Signal in
Upper Limit
Lower Limit
Error in (no error)
Limit Test Config

Failures
Test Passed?
Test Results
Output Values
Error Out
Clearance

Peak Searching and Fundamental Detection

Spectra
Peak Search Settings
Spectrum Info
Error in (no error)

Peaks
Number of Peaks
Unit Labels
Error Out

Desired Units
Scaled Signal [EU]
Maximum Harmonic
Harmonics to Visualize
Error in (no error)
Expected Fundamental Freque...

Actual Maximum Harmonic
Unit Labels
Complex Spectrum
THD
Harmonic Components
Error Out
Detected Fundamental Amplitude
Detected Fundamental Freque...

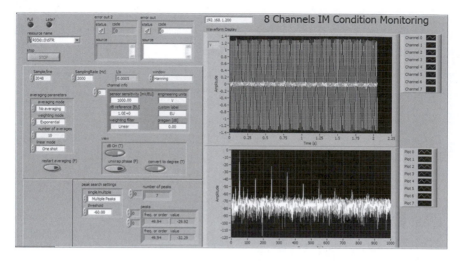

FIGURE 6.10 User interface of the CompactRIO-based condition monitoring system.

the status of the motor to a custom-design indicator module through SPI triggering technique, and it captures data and transfers it through the PCI bus to the real-time host controller. Several techniques are implemented for the host synchronization such as polling, interrupt request, and direct memory access. These functions are programmed through the LabVIEW FPGA environment.

Resources

The CompactRIO platform has a small rugged, modular, scalable, embedded open architecture system consisting of an FPGA, a real-time processor, and isolated I/O devices. Since this device has industrial class certification (class I and div2) for hazardous locations or potentially explosive, high temperatures, and 50 g shock capability, it is a good candidate to be used in an industrial environment.

The CompactRIO Module Development Kit (MDK) is also required to expand the capability of the CompactRIO to meet the unique needs of the application, the indicator module. The MDK licenses users to develop and manufacture their own CompactRIO modules. This development kit comprises 10 empty module shells, development software, and a variety of header connectors, such as D-Sub, BNC, and screw terminals. CompactRIO MDK also presents the hardware requirements and recommendations, EEPROM format, and development of VI drivers for the I/O modules.

Following the arrangement of the CompactRIO base unit mentioned previously, three main modules are also needed for the CompactRIO-based condition monitoring system: (1) a real-time embedded controller module (cRIO-9004); (2) a reconfigurable chassis containing user programmable FPGA (cRIO-9104); and (3) an analog input module (NI 9201). Another important module is the custom-made CompactRIO indicator module.

An analog input module NI 9201 (C-series, eight channels, 12-bit SAR ADC, 500 kS/s) is used to capture vibration, current, voltage, and flux leakage signals using suitable sensors and signal conditioning devices. The total input channels meet the requirement of sensory signals that are employed for the condition monitoring system. This module is suitable for condition monitoring applications because it has eight analog input channels to capture 10.8 V maximum signals. In addition, this module also contains channel-to-earth ground isolation barrier up to 100 V, isolated analog-to-digital converter (ADC), and noise immunity system, which are desirable features in signal measurements in condition monitoring.

Table 6.4 lists the sensor types, signals, and associated bandwidths and some descriptions of their positions.

TABLE 6.4 Sensor Parameters and Positions

Monitoring Technique	Signals	Sensors	FFT Bandwidth	Position
Vibration	Acceleration velocity displacement	Accelerometer	1–2 g 10 Hz to 1 KHz	Motor drive end and nondrive end: axial, radial-horizontal, radial-vertical
Stator current	Current	Current transducers: Hall-Effect type or Rogowski coils	1,000 Hz, 20th harmonics	Clipped to one, two, or three phase supply
Stator voltage	Voltage	Voltage transformers, differential isolation amplifiers	1,000Hz, 20th harmonics	Clipped to one, two, or three phase supply
Flux leakage monitoring	Magnetic field	Search coil or Hall sensors (external or internal)	mV range, 1,000 Hz, 20th harmonics	Nondrive end or motor outboard around the shaft

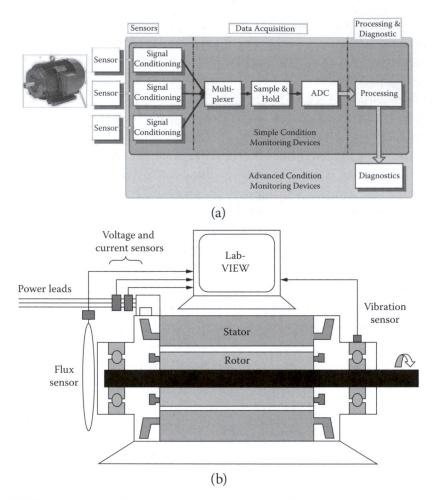

FIGURE 6.11   Block diagram of an advanced condition monitoring system (a) and the LabVIEW based structure including a PC and associated sensors (b).

PC-Based Online Condition Monitoring System

Figure 6.11 shows data acquisition and processing main tasks in an advanced condition monitoring system. The simple LabVIEW-based condition monitoring device may not cover all tasks shown in the figure, such as the diagnostic function, but it contains a data acquisition task with a multiplexer and a sample and hold function, which can be built from a high-speed amplifier and an ADC.

Note that the motor test setup shown in Figure 6.11a is similar to the CompactRIO-based system given earlier. In summary, the following variables are also measured on the PC-based system:

- Two line voltage signals (the third line voltage was calculated).

- Two phase stator current signals (the third phase was calculated assuming a three-wire connection).

- Three vibration signals, which were captured at three different positions: driving end horizontal (DEH) position; driving end vertical (DEV) position; and nondriving end horizontal (NDEH) position.

- One axial leakage flux signal, which is the magnetic flux that radiates out of the motor frame.

As also indicated previously, the position of the search coil was located around the motor housing near the nondrive end for easier access. It should be emphasized here that the orientation of the search coil is important as it can affect the induced voltage and the level of the fault frequency. The method of mounting the vibration sensor also influences the frequency response of measured signal. The ideal method is to screw mount the vibration sensor to the motor through drilled taps on the motor's housing (Figure 6.12).

Each of the analog signals from the sensors was passed to a variable gain amplifier and a low pass filter. The variable gain amplifier has eight channels, and each channel is able to adjust the gain (i.e., x1, x10, and x100) individually to amplify a weak signal. The low pass filter is an eight-channel eighth-order Butterworth analog filter with a selectable cutoff

FIGURE 6.12 Photograph of the motor test setup showing the positions of the vibration sensors.

frequency of either 100 Hz or 2 kHz. The cutoff frequency was chosen to satisfy the Nyquist theorem, where the sampling rate must be at least twice the highest frequency present in the signal. The low pass filter was used to remove any unwanted high-frequency components that could otherwise cause aliasing issues. The filtered signals were then passed to ADCs, which are part of the data acquisition hardware, to obtain a set of digital data. The data acquisition hardware is a plug-in card from National Instruments (NI-PCI-6110, 12-bit, 5 MHz, simultaneous sampling). A simple flow diagram of the condition monitoring system is shown in Figure 6.16.

Ideally, all sensors should be sampled at a sampling rate and sampling time as high as possible, but these options are restricted by the hardware limitations. The sensors are sampled simultaneously with two different sampling rates and two different sampling times, which were chosen as follows:

- Low-frequency measurement at 400 Hz sampling frequency (which gives a Nyquist frequency of 200 Hz) with a sampling time of 100 s, which allows very high-resolution frequency analysis with a lower bandwidth (40,000 data points with 0.01 Hz resolution).

- High-frequency measurement at 8000 Hz sampling frequency with a sampling time of 5 seconds, which allows high-frequency analysis with a lower frequency resolution (40,000 data points with 0.2 Hz resolution).

A summary of sensor signal sampling information is given in Table 6.5. It should be emphasized here that these two sampling (and cutoff) frequencies were particularly chosen because most of the low-frequency components that are of interest usually lie between 0 and 100 Hz and the high-frequency components that are of interest usually lie below 2 kHz.

TABLE 6.5 Summary of Sensor Signal Sampling Information

	Low-Frequency Sampling	High-Frequency Sampling
Sampling Frequency	400 Hz	8 kHz
Total Sampling Time (T)	100 sec	5 sec
Frequency Resolution $\Delta f = (1/T)$	0.01 Hz	0.2 Hz
Total number of points in record $L_R = fs\Delta f$	40,000	40,000
Nyquist frequency	200 Hz	4 kHz
Cutoff frequency	100 Hz	2 kHz

Furthermore, the 40,000 data points were chosen because of the limitations of the hardware processing power.

The custom LabVIEW program is used to record the sensor signals for the three measurement cases in a slip-ring induction machine: Healthy Test, Faulty Test 1 (rotor fault by adding a rotor resistor), and Faulty Test 2 (supply imbalance by adding a stator resistor to one of the motor's phases). The tests are repeated to capture data under both low- and high-frequency sampling settings. The front panels during the tests include:

"Setting panel" where the "File Path," "Motor's name," "Motor's fault," and "File Name" can be entered (Figure 6.13)

"Waveform panel" (Figure 6.13) where the current, voltage, flux, and vibration signals are shown. Pressing "Settings" on this panel resets the saving parameters.

"Start Capture" and "Sampling frequency" panels (Figure 6.14) display a panel to set the sampling frequency (to 400 Hz for low-frequency data and to 8000 Hz for high-frequency data) and then press "Start Capturing" to capture the signals.

A sample file that shows the captured signals is given in Figure 6.15. The test number is automatically appended to the file name to differentiate the multiple files. All other phase signals and their spectrums can also be checked by clicking the corresponding buttons located under the graphs.

The spectrum analysis of the measured signals is performed using the "Spectrum system" VI (Figure 6.16).

To view the frequency spectrum of a signal, the appropriate type of spectrum on the top right of the screen (i.e., Current, Voltage, Flux, Vibration) the "OK" button should be pressed. A full spectrum and a zoomed spectrum of the selected signals are displayed on the left- and right-hand sides of the associated spectrum front panels in Figure 6.17.

Note that the horizontal and vertical axis scaling of the graphs can be changed in the panels of Figure 6.17 by selecting and editing the beginning and ending axis labels, which allows the user to zoom in on areas of interest for accurate determination of fault frequencies and amplitudes. Also note that a list of peaks is displayed in the center of the screen. If there are more peaks detected than can be displayed, the user can scroll up and down the list with the scroll bars to the left of the first peak displayed. The threshold for the peak detection can be set by dragging the

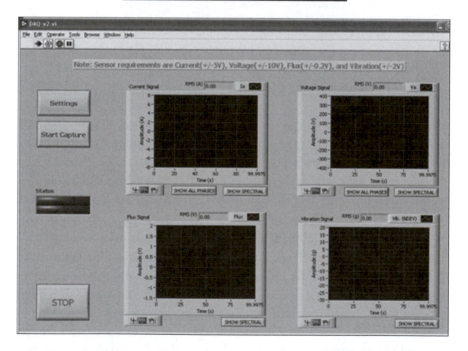

FIGURE 6.13 The front panels: "Setting panel" (top) and "Waveforms front panel" (bottom).

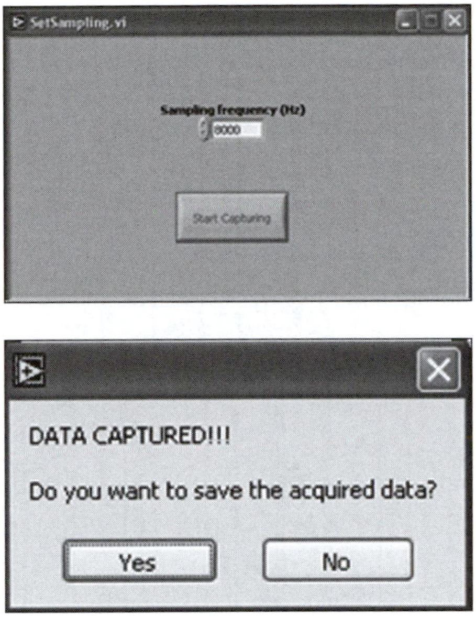

FIGURE 6.14 "Sampling frequency panel" (top) and "Saving panel" (bottom).

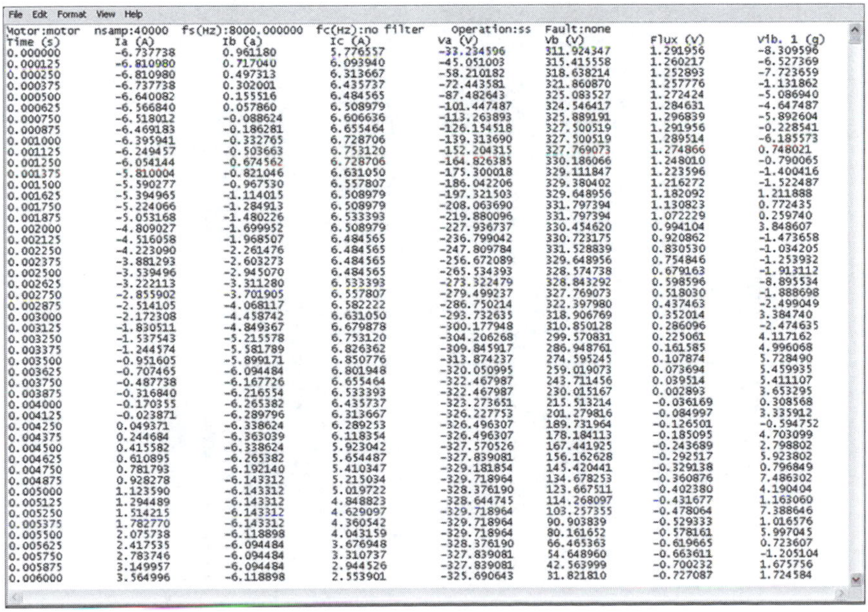

File Edit Format View Help							
Motor:motor	nsamp:40000	fs(Hz):8000.000000	fc(Hz):no filter	Operation:ss	Fault:none		
Time (s)	Ia (A)	Ib (a)	Ic (A)	va (V)	Vb (V)	Flux (V)	Vib. 1 (g)
0.000000	-6.737738	0.961180	5.776557	-33.234596	311.924347	1.291956	-8.309596
0.000125	-6.810980	0.717040	6.093940	-45.051003	315.415558	1.260217	-6.527369
0.000250	-6.810980	0.497313	6.313667	-58.210182	318.638214	1.252893	-7.723659
0.000375	-6.737738	0.302001	6.435737	-72.443581	321.860870	1.257776	-1.131862
0.000500	-6.640082	0.155516	6.484565	-87.482643	325.083527	1.272424	-5.086940
0.000625	-6.566840	0.057860	6.508979	-101.447487	324.546417	1.284631	-4.647487
0.000750	-6.518012	-0.088624	6.606636	-113.263893	325.889191	1.296839	-5.892604
0.000875	-6.469183	-0.186281	6.655464	-126.154518	327.500519	1.291956	-0.228541
0.001000	-6.395941	-0.332765	6.728706	-139.313690	327.500519	1.289514	-6.185573
0.001125	-6.249457	-0.503663	6.753120	-152.204315	327.769073	1.274866	0.748021
0.001250	-6.054144	-0.674562	6.728706	-164.826385	330.186066	1.248010	-0.790065
0.001375	-5.810004	-0.821046	6.631050	-175.300018	329.111847	1.223596	-1.400416
0.001500	-5.590277	-0.967530	6.557807	-186.042206	329.380402	1.216272	-1.522487
0.001625	-5.394965	-1.114015	6.508979	-197.321503	329.648956	1.182092	1.211888
0.001750	-5.224066	-1.284913	6.508979	-208.063690	331.797394	1.130823	0.772435
0.001875	-5.053168	-1.480226	6.533393	-219.880096	331.797394	1.072229	0.259740
0.002000	-4.809027	-1.699952	6.508979	-227.936737	330.454620	0.994104	3.848607
0.002125	-4.516058	-1.968507	6.484565	-236.799042	330.723175	0.920862	-1.473658
0.002250	-4.223090	-2.261476	6.484565	-247.809784	331.528839	0.830530	-1.034205
0.002375	-3.881293	-2.603273	6.484565	-256.672089	329.648956	0.754846	-1.253932
0.002500	-3.539496	-2.945070	6.484565	-265.534393	328.574738	0.679163	-1.913112
0.002625	-3.222113	-3.311280	6.533393	-273.322479	328.843292	0.598596	-8.895534
0.002750	-2.855902	-3.701905	6.557807	-279.499237	327.769073	0.518030	-1.888698
0.002875	-2.514105	-4.068117	6.582222	-286.750214	322.397980	0.437463	-2.499049
0.003000	-2.172308	-4.458742	6.631050	-293.732635	318.906769	0.352014	3.384740
0.003125	-1.830511	-4.849367	6.679878	-300.177948	310.850128	0.286096	-2.474635
0.003250	-1.537543	-5.215578	6.753120	-304.206268	299.570831	0.225061	4.117162
0.003375	-1.244574	-5.581789	6.826362	-309.841917	286.948761	0.161585	4.996068
0.003500	-0.951605	-5.899171	6.850776	-313.874237	274.595245	0.107874	5.728490
0.003625	-0.707465	-6.094484	6.801948	-320.050995	259.010093	0.073694	5.459935
0.003750	-0.487738	-6.167726	6.655464	-322.467987	243.711456	0.039514	5.411107
0.003875	-0.316840	-6.216534	6.533393	-322.467987	230.015167	0.002893	3.653295
0.004000	-0.170355	-6.265382	6.435737	-323.273651	215.513214	-0.036169	0.308568
0.004125	-0.023871	-6.289796	6.313667	-326.227753	201.279816	-0.084997	3.335912
0.004250	0.049371	-6.338624	6.289253	-326.496307	189.731964	-0.126501	-0.594752
0.004375	0.244684	-6.363039	6.118354	-326.496307	178.184113	-0.185095	4.703099
0.004500	0.415582	-6.338624	5.923042	-327.570526	167.441925	-0.243689	2.798802
0.004625	0.610895	-6.265382	5.654487	-327.839081	156.162628	-0.292517	5.923802
0.004750	0.781793	-6.192140	5.410347	-329.181854	145.420441	-0.329138	0.796849
0.004875	0.928278	-6.143312	5.215034	-329.718964	134.678253	-0.360876	7.486302
0.005000	1.123590	-6.143312	5.019722	-328.376190	123.667511	-0.402380	4.190404
0.005125	1.294489	-6.143312	4.848823	-328.644745	114.268097	-0.431677	1.163060
0.005250	1.514215	-6.143312	4.629097	-329.718964	103.257355	-0.478064	7.388646
0.005375	1.782770	-6.143312	4.360542	-329.718964	90.903839	-0.529333	1.016576
0.005500	2.075738	-6.118898	4.043159	-329.718964	80.161652	-0.578161	5.997045
0.005625	2.417535	-6.094484	3.676948	-328.376190	66.465363	-0.619665	0.723607
0.005750	2.783746	-6.094484	3.310737	-327.839081	54.648960	-0.663611	-1.205104
0.005875	3.149957	-6.094484	2.944526	-327.839081	42.563699	-0.700232	1.675756
0.006000	3.564996	-6.118898	2.553901	-325.690643	31.821810	-0.727087	1.724584

FIGURE 6.15 Example of data saved.

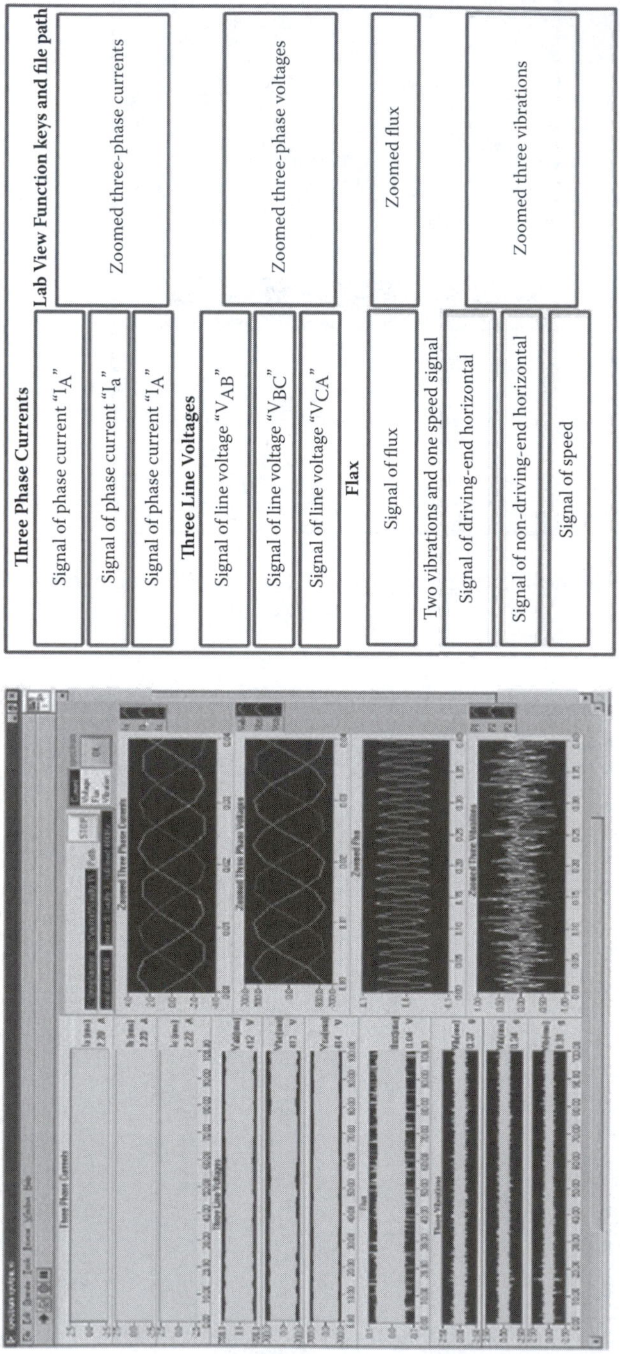

Three Phase Currents	Lab View Function keys and file path
Signal of phase current "I_A"	
Signal of phase current "I_a"	Zoomed three-phase currents
Signal of phase current "I_a"	
Three Line Voltages	
Signal of line voltage "V_{AB}"	
Signal of line voltage "V_{BC}"	Zoomed three-phase voltages
Signal of line voltage "V_{CA}"	
Flax	
Signal of flux	Zoomed flux
Two vibrations and one speed signal	
Signal of driving-end horizontal	
Signal of non-driving-end horizontal	Zoomed three vibrations
Signal of speed	

FIGURE 6.16   The front panel of Spectrum Analysis program (left) and a brief user guide (right).

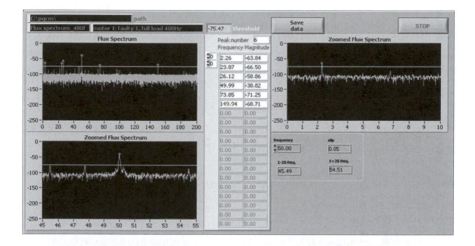

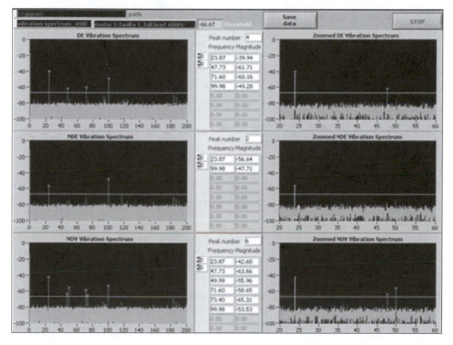

FIGURE 6.17 Sample panels of the flux frequency spectrum (top) and the three line voltage frequency spectrums (bottom).

horizontal line on the graph with the mouse. However, some peaks in the fault diagnosis of induction motors may not be automatically detected and hence may have to be searched manually. The peaks listed can be saved in a spreadsheet file by choosing an appropriate path and file name first and by updating the comment field (top left of screen).

Resources

The sensor specifications used in this application are similar to the CompactRIO solution explained previously: eight sensors in total (two voltage sensors, two current sensors, two vibration sensors, flux sensor, and a speed sensor).

The data acquisition system used in this study consists of two four-channel plug-in cards from National Instruments (NI-PCI-6110, 12-bit, 5 MS/sec, simultaneous sampling), anti-aliasing filter.

Real-Time Solar Array Monitoring System in Remote Areas

In solar-based renewable energy systems, a number of single photovoltaic (PV) cells are usually configured in series or in parallel to form an array to increase voltage and current ratings. The PV arrays are also covered with some form of protection material and are assembled in a frame to increase their reliability against environmental factors and to simplify installation. As these arrays operate in various regions and environmental conditions, it is necessary to obtain their performance characteristics and to be able to accurately predict their energy output. Therefore, rapid improvements in the manufacture of solar panels demand standardized tests to evaluate and compare solar panel performance. Because the energy output in a PV array depends on the level of solar insolation and the cell temperature (both are unpredictable), it is critical to develop an accurate and repeatable measurement system. The VI described in this section aims to offer a LabVIEW-based measurement system to obtain the characteristics of PV arrays.

Standard parameters for PV panel comparison include I_{SC} (short-circuit current), V_{OC} (open-circuit voltage), P_{max}, V_{max}, I_{max} (maximum power-point power, voltage, and current), FF (fill factor), and η (efficiency), all of which can vary with T (temperature) and G (insolation) for a given solar panel. To measure these parameters, a single inexpensive unit was designed in this LabVIEW application. This system is required to be robust, to be capable of handling harsh environments like the Australian outback, and to handle a wide range of solar panels or arrays of solar panels. The application reported here concentrates on the design of a maximum power point tracker (MPPT) that can measure P_{max}, V_{max}, and I_{max} and maintain arrays at the maximum power point to aid measurement of the other parameters.

In a remote area PV array monitoring system, custom software is required to work with custom-made hardware (Figure 6.18). Although the hardware system is relatively easy to build, it is critical to integrate it to the

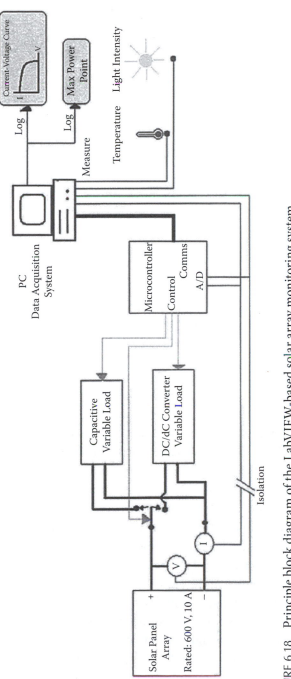

FIGURE 6.18 Principle block diagram of the LabVIEW-based solar array monitoring system.

software while considering limitations such as pulse width modulation (PWM). In this application the software is used to

- Control the hardware (static switches, LEDs)

- Run current-voltage (I-V) and MPPT tests

- Measure voltage, current, and temperature

- Save measured data

- Automate the procedures for repeating tests

The software component of the entire system has two main sections: an embedded microcontroller to control hardware, and a data acquisition VI to measure various parameters and to save. The microcontroller also acts as a control interface between the switching hardware and the data acquisition system and is responsible for the real-time control of a capacitive load to generate I-V characteristics of a given PV array. It was aimed at placing the control sequencing onto the microprocessor to increase the speed by allowing the data acquisition system to solely handle the acquisition of data. The communication among the microcontroller and the switching hardware and the data acquisition system is done via digital I/O lines.

The control variables of the I-V software are (1) the solar panel/capacitor state, which provides the connection between the panel and the capacitor load bank; (2) the capacitor/resistor state (that provides the connection between the capacitor load bank and the discharge resistor); and (3) the trigger state (to instigate the data acquisition). The flow diagram of the I-V section of the microprocessor software is shown in Figure 6.19.

Upon exiting this section, the software control is handed back over the main software loop, which determines when the MPPT is achieved. When the MPPT digital line output is detected within the mail loop, the microcontroller software transfers control to the MPPT program, which initializes variables, turns on PWM output, and then turns on the ADC. After initialization is complete, the MPPT program enters the "Tracking loop" until the maximum power point (MPP) is found or the test is canceled. The "Tracking loop" contains a number of sequences: determination of MPP, measuring voltage and current, averaging voltage and current, calculating power, determining change in power, comparing the level power respect to a threshold. Upon quitting the "Tracking loop," if the MPP flag has been set by the program, then it reports this to the LabVIEW-based

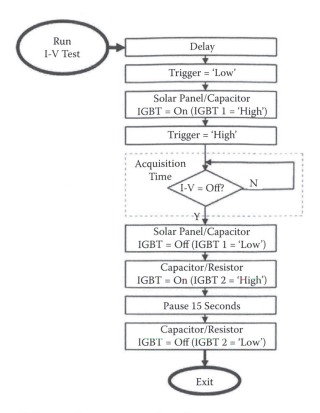

FIGURE 6.19   I-V test software section flow diagram.

software. Automated acquisition is used to schedule and automatically run I-V and MPPT tests up to 15 times in a day.

As shown in the block diagram in Figure 6.20, the automated acquisition block diagram is surrounded by a case structure, which starts when the "Start Automated Acquisition" button is true. The automated acquisition continues (the While Loop continues) until either the stop button is pressed or an error occurs. The test times for the I-V and MPPT tests are fed into the scheduler subVI. If a scheduled I-V test time matches the current system time, then the scheduler returns a true value for the I-V test. An I-V test is then executed (with corresponding temperatures and light intensity also measured) using the I-V subVI. If a scheduled MPPT test time matches the current system time, then the scheduler returns a true value for MPPT test. A MPPT test is then executed using the MPPT subVI.

The "I-V Test Configuration Utility" (Figure 6.21) is used to configure various parameters regarding the I-V test, which include "Save Path," "Acquisition Time," "Number of Pre-trigger Scans," "Number of Scans to

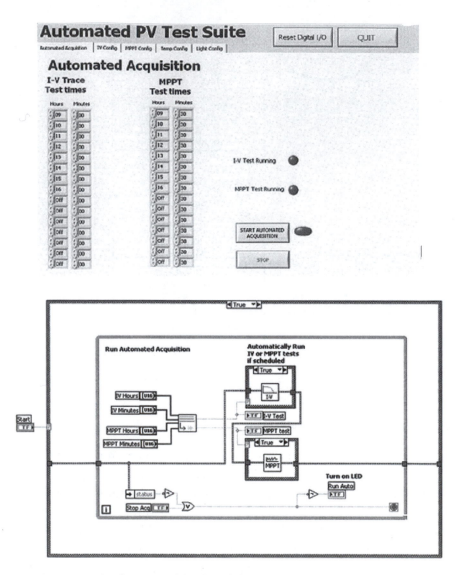

FIGURE 6.20 The front panel (top) and the diagram of the automated acquisition (bottom).

Read at a Time," "Scan Rate" per channel, and "Averaging." This utility allows the user to run a single I-V test on demand. All the measured values of voltage and current are plotted on their respective charts, which allow the user to determine if the values configured will be suitable to obtain a successful I-V test result. All the values are recorded into a spreadsheet file. As can be seen in the figure, the I-V test configuration utility block diagram is also surrounded by a case structure, which runs when the

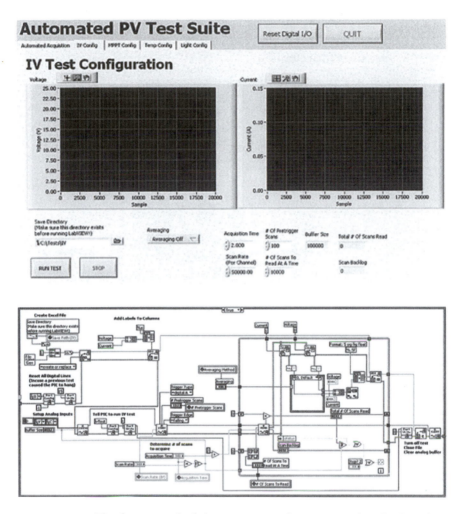

FIGURE 6.21 The front panel of the I-V test configuration utility (top) and its block diagram (bottom).

"Run Test" switch is true. The file name of the spreadsheet file is generated automatically and depends on the current date and time.

Data acquisition is triggered, and a circular buffer is used. Data acquisition starts when LabVIEW is triggered by the microcontroller. The LabVIEW program triggers the microcontroller to run the MPPT test in configuration mode (Figure 6.22). Therefore, at the end of every iteration (i.e., where the MPPT algorithm adjusts the power point by one step) the microcontroller triggers LabVIEW to measure the values of voltage and current from which power is calculated. Then LabVIEW waits to measure the next value of power. The values are saved into a spreadsheet file.

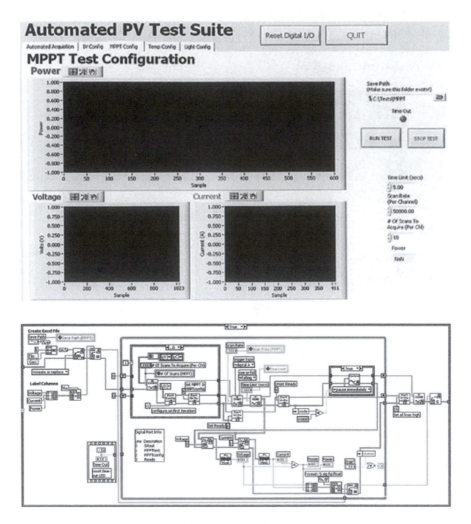

FIGURE 6.22   MPPT Test Configuration Utility.

Similar to the previous modules, the Temperature Test Configuration Utility (Figure 6.23) is used to configure various parameters regarding the measurement of the ambient, panel, and PV cell temperatures. In this section of the test, three temperature measurements are taken and displayed on their respective graphs and their average values are displayed in the indicators.

The Light Intensity Test Configuration Utility (Figure 6.24) is used to make a single light intensity measurement, which is also displayed on the respective graph and the indictor of the front panel.

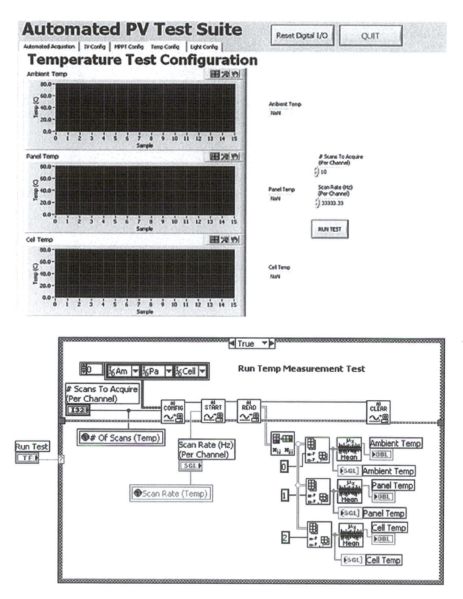

FIGURE 6.23   Temperature Test Configuration Utility.

A number of subVIs are developed in this program including "Scheduler" to read the arrays of the I-V and MPPT test times and "I-V Test" to perform very similar operations as in the "I-V Test Configuration Utility" (Figure 6.25).

Resources
The DAQ card system used in this application should perform a number of tasks in a remote environment: providing the interface with the microcontroller,

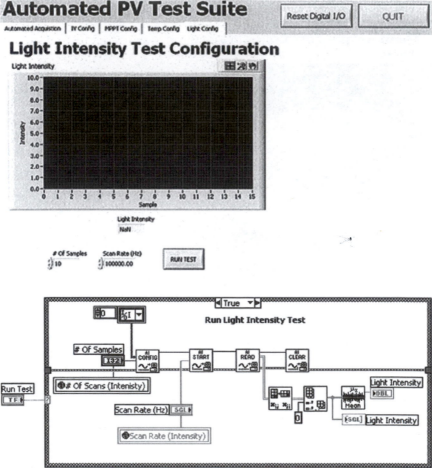

FIGURE 6.24 Light Intensity Test Configuration Utility front panel and block diagram.

taking measurements, capturing data to a file, and autonomously running I-V and MPPT tests throughout the day. Therefore, a CompactRIO-based system or a laptop or DAQ card based system can be suitable for this application. The DAQ card should have a minimum of four analog input channels and digital outputs to control the microcontroller. The system tested in this application is illustrated in Figure 6.18 and has the following features:

$$V_{OC} = 10 \text{ to } 600 \text{ V voltage range}$$

$$I_{SC} = 10 \text{ mA to } 12 \text{ A current range}$$

Temperature range: 0–100°C

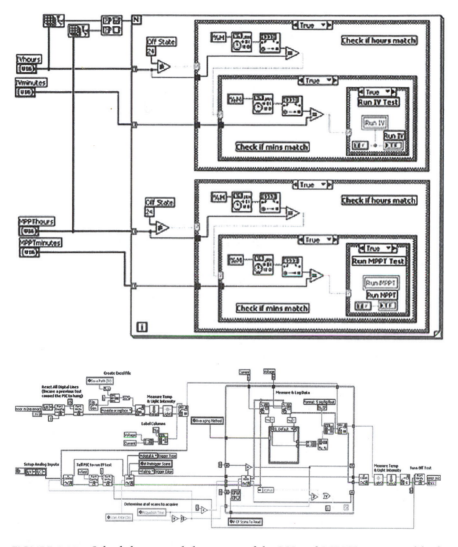

FIGURE 6.25 Scheduler to read the arrays of the I-V and MPPT test times block diagrams.

Resolution: 12 bit

Accuracy: 2% of actual values

Voltage and current sensors should be able to measure these ranges accurately. The accompanying DAQ card used in the development of this application was a relatively early model, National Instruments AT MIO-16E-10. Therefore, it can be concluded that any low-level DAQ card is capable of performing the required task in this LabVIEW application.

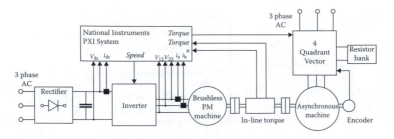

FIGURE 6.26   The block diagram of the automated dynamometer test setup.

Automated Dynamometer Test Setup

Electrical motor tests are some of the most time-consuming and cumbersome tests due to their nonlinear operating characteristics and the difficulties of accurate and repeatable loading. The developments in computer-aided data acquisition systems and electronically controlled motor drives, however, have made such tests relatively easy to implement.

The block diagram of the method presented in this section is described in Figure 6.26. As shown in the figure, the machine under test is attached to a load machine that can operate in four quadrants using a vector drive. In addition, both the motor under test and the loading machine are mechanically coupled via an in-line torque transducer that also includes speed sensor. Furthermore, one DC link voltage and one DC link current, two line voltages, two line currents, instantaneous torque, and speed signals are measured via a PXI system. The PXI system is used to accommodate custom VIs to control both the speed and the torque of the motor and the load.

Basic operational links of the system are shown in Figure 6.27. In the following figures, the front panel snapshots indicate the operational features of the test setup. The main front panel, illustrated in Figure 6.28, provides a link to the three primary subpanels: Setting Panel, Power Analyzer, and Auto Testing. The main front panel offers two additional functions to perform motor loading or generator loading tests. Note that the block diagram of the test setup is provided on the front panel to imitate conventional instrumentation. The measured efficiencies of the subsections of the system under test are also included. In addition, the instaneous signals can be displayed in the program (Figure 6.29).

In this system, an automatic report generation was also accommodated using the built-in features of LabVIEW programming. A sample structure of such a report follows and should be used as a reference; it contains a selected number of graphs to demonstrate the capabilities of the test system.

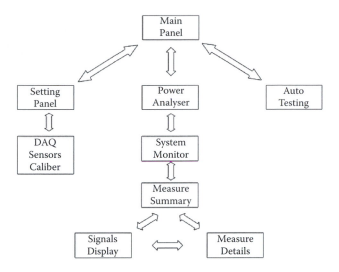

FIGURE 6.27   Operational links of the automated test system.

SAMPLE MACHINE TEST REPORT

STARTING TEST

The starting test results of the machine SN: IM1 (1.2 kW) is given as follows. The test was conducted on February 22, 2007, at 9:25 a.m. The total starting time of the machine is measured as 1.6 s, and the maximum value of the current during starting is measured as 3.04 A, which occurred at 430.50 ms after starting (Figure 6.30).

STEADY-STATE TEST

The steady-state test of the machine was done on February 22, 2007, at 9:37 a.m. During the test, the machine was loaded from no load to 100% (1.2 kW). During the automated tests, speed versus input power, torque, shaft power, and efficiency characteristics of the machine were obtained.

VIBRATION TESTS

The Vibration test of the machine was performed on February 22, 2007, at 10:45 a.m. Four accelerometers were used and were attached to the motor housings as Sensor 1 (Drive End Vertical), Sensor 2 (Nondrive End Vertical), Sensor 3 (Drive-End Horizontal), and Sensor 4 (Nondrive End Horizontal). The measured vibration signals from these sensors are shown in Figures 6.31 through 6.34.

Figure 6.35 shows the frequency components of the measured vibrations waveforms that have amplitudes above a threshold of 0.01.

STEADY-STATE TEST

The Temperature test was conducted on February 22, 2007, at 11:05 a.m., and the measured temperature patterns from eight sensors are shown in Figure 6.36.

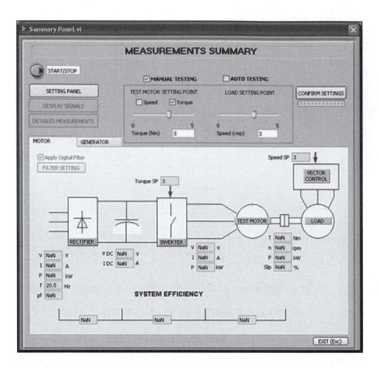

FIGURE 6.28 The main front panel (top) and measurement summary front panel (bottom).

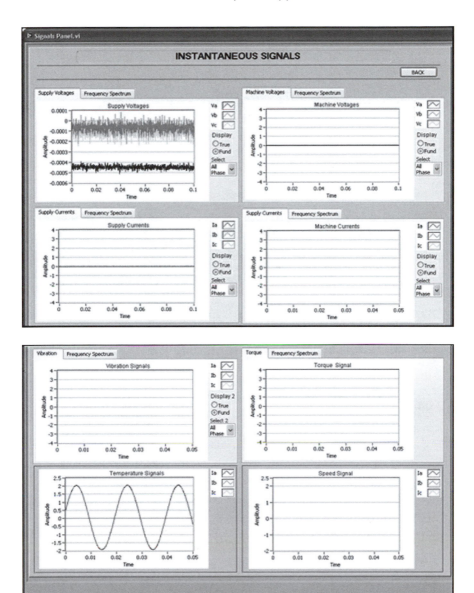

FIGURE 6.29   Signal display panels.

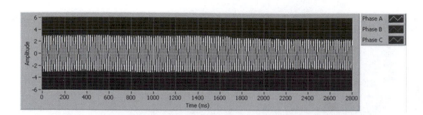

FIGURE 6.30   Starting current signals.

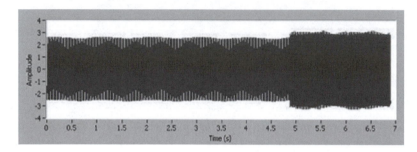

FIGURE 6.31   Vibration signal—sensor 1.

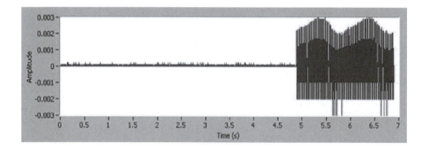

FIGURE 6.32   Vibration signal—sensor 2.

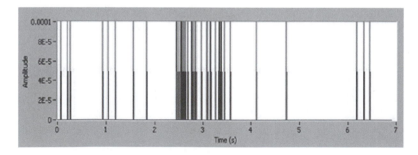

FIGURE 6.33   Vibration signal—sensor 3.

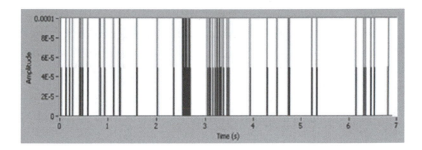

FIGURE 6.34   Vibration signal—sensor 4.

Sensor I		Sensor II		Sensor III		Sensor IV	
Freq (Hz)	Amplitude	Freq (Hz)	Amplitude	Freq (Hz)	Amplitude	Freq (Hz)	Amplitude
1.08	-29.36						
48.32	-29.52						
48.85	-27.90						
49.34	-25.58						
49.89	-22.53						
51.07	4.07						
52.69	-25.65						
53.18	-27.96						
53.70	-29.81						
153.09	-18.66						
255.15	-23.12						
357.22	-26.01						
459.29	-28.15						
561.34	-30.02						

FIGURE 6.35   The frequency components of the measured vibrations waveforms.

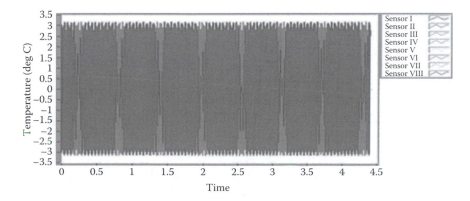

FIGURE 6.36   Measured temperature.

Resources

- National Instruments PXI system, with a DAQ card that has a minimum of 10 channels AIO (plus additional channels for temperature measurements)

- Voltage and current sensors with suitable ratings to match the ratings of the electrical machines under tests and their inverters

Electrical Machines Tests

This test system is also developed to simplify the testing procedures of DC electrical machines using the benefits of LabVIEW and associated data acquisition hardware. In this application, the software starts by displaying the main panel, which shows the main menu of the test (top of Figure 6.37). Prior to starting the measurement process, the specifications of the machine are entered by clicking on the top button on the panel to activate the Machine Specification Panel (bottom of Figure 6.37).

Once the data are entered and the specifications are saved by clicking on the OK button, an inquiring dialog box can appear to specify whether the measurements are done in "cold" condition or not. Selecting No (not in cold condition!) returns the software to the main menu. Selecting Yes opens the Cold Measurement Panel. The data in this panel have to be entered manually. When the cold measurements are entered and saved, the main menu is displayed for starting the tests. To be able to start the tests, a directory name should also be specified on the bottom of the Main Panel. The directory name can be either typed in the specified space or selected by clicking on the directory icon. By clicking on the directory icon, a navigating dialog box appears showing all the directories in the host computer.

Once the machine data are entered and the directory is selected successfully, the tests can start as follows. As per any test in this chapter, a zero calibration is done after connecting the sensors to the DAQ card and prior to connecting them to the devices under test. This can be done by clicking on the Calibration button on the top left of the panel. After a few seconds, a dialog box will occur to confirm the completion of the calibration. After finishing the calibration, the acquisition settings of the starting signals can be adjusted from the Settings button.

The DC measurement panel (top of Figure 6.38) has four tabs, outlined in the following sections.

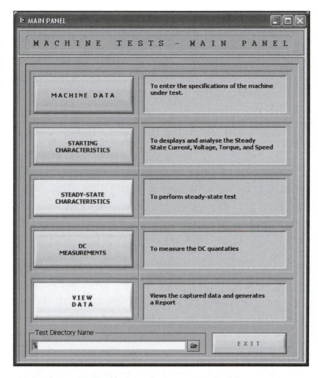

FIGURE 6.37   The main front panel (top) and the machine specifications panel (bottom) of the electrical machine test.

DC System Layout Tab

The first tab is for displaying the layout of the system showing the online measurements of the DC quantities. However, before these quantities can be displayed, the calibration of the system has to be done in this tab. The Calibration is done in a similar way to the AC tests. Also, before starting the measurements, the settings of the acquisition and the machine information can be adjusted by accessing the "SETTINGS" button (bottom of Figure 6.38).

After adjusting and saving the required settings, the measurements can be started using the START button. During this step, online measurements are displayed and can be saved in a spreadsheet by clicking "Save Displayed Data." This includes a substep where the name and path of the spreadsheet file are shown.

It should be reported here that every time the save button is pushed all the displayed information will be appended to the same file. The number of the saved data is indicated in the "Test Number" indicator. A sample spreadsheet file with four sample measurements is shown in Figure 6.39.

Commutation Test Tab

The second tab in the panel is for the Commutation test (Figure 6.40). After filling in the information on the top of the panel, the field current is adjusted for filling the tables given. Online measurements of the field current and armature voltage and current are displayed in the middle of the panel. Once the required field current is displayed on the panel, the data can be captured and displayed in the left-hand table in the figure. Provided that the field current is not changed, every time the capture button is pushed a new measurement will be entered into the same table. Note that a percentage of variation can be ignored based on the value entered in the "Field current tolerance (%)" space. To fill the right-hand table in the figure, the field current has to be changed to a different value. Returning the field current to the first value will fill in again the first table. The filling of either the top or the bottom row tables is controlled by the direction radio button provided on the same front panel. After filling the four tables, the over-speed test can be performed by activating the test on the panel.

The rotating speed of the machine is displayed on the bottom right side of the panel, which can be captured and displayed in the custom report.

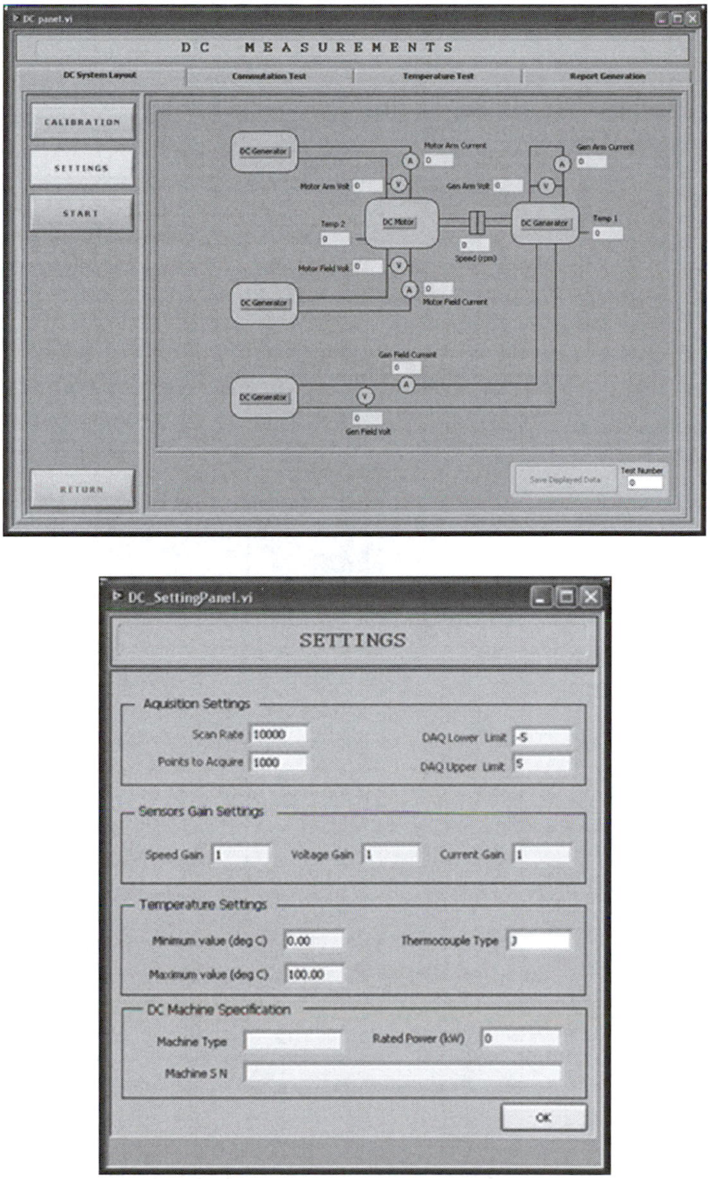

FIGURE 6.38 The main panel (top) showing the layout of the measuring system, and the Settings front panel (bottom).

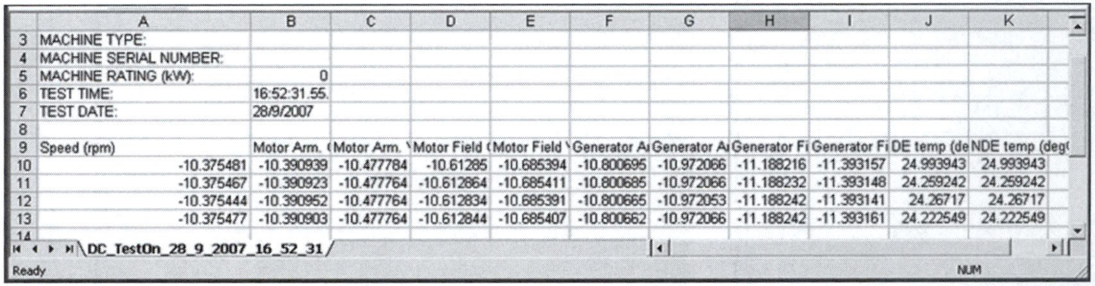

	A	B	C	D	E	F	G	H	I	J	K
3	MACHINE TYPE:										
4	MACHINE SERIAL NUMBER:										
5	MACHINE RATING (kW):	0									
6	TEST TIME:	16:52:31.55.									
7	TEST DATE:	28/9/2007									
8											
9	Speed (rpm)	Motor Arm. '	Motor Arm. '	Motor Field '	Motor Field '	Generator A	Generator A	Generator Fi	Generator Fi	DE temp (de	NDE temp (deg
10		-10.375481	-10.390939	-10.477784	-10.61285	-10.685394	-10.800695	-10.972066	-11.188216	-11.393157	24.993943	24.993943
11		-10.375467	-10.390923	-10.477764	-10.612864	-10.685411	-10.800695	-10.972066	-11.188232	-11.393148	24.259242	24.259242
12		-10.375444	-10.390952	-10.477764	-10.612834	-10.685391	-10.800665	-10.972053	-11.188242	-11.393141	24.26717	24.26717
13		-10.375477	-10.390903	-10.477764	-10.612844	-10.685407	-10.800662	-10.972066	-11.188242	-11.393161	24.222549	24.222549
14											

DE_TestOn_28_9_2007_16_52_31

Ready   NUM

FIGURE 6.39 A sample spreadsheet file with four sample measurements.

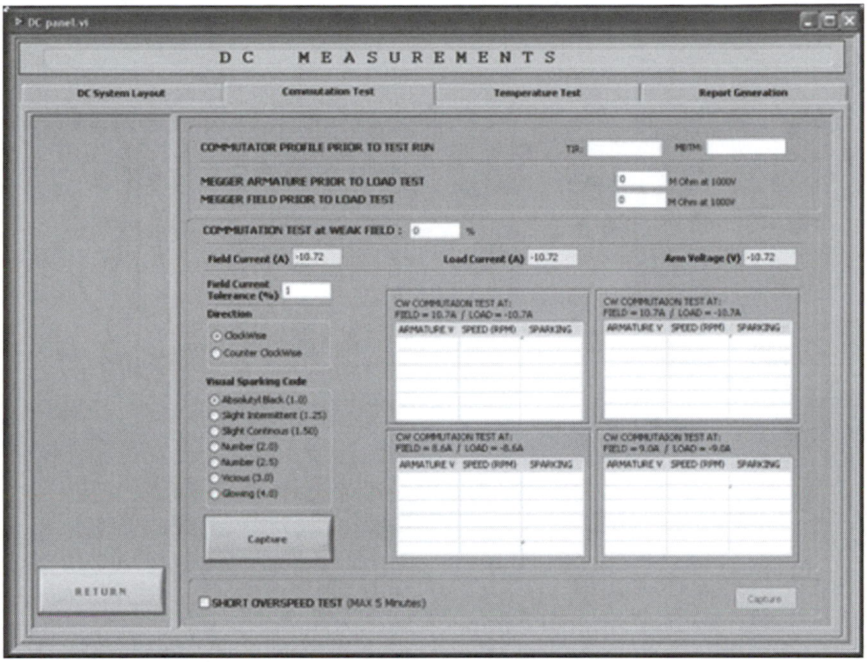

FIGURE 6.40   Commutation test front panel.

Temperature Test Tab

The temperature test panel (Figure 6.41) is similar to the temperature test panel of the AC test. However, in this panel, the operator can select whether the measurements will be done while the motor is running or while it is stationary. The measurements are entered manually on the bottom of the panel when the machine is stopped, whereas the software will capture the measurements if the machine is running. When the machine is running, the "Sampling Interval" and the "Total Test Time" have to be entered to specify the period between each two measurements and the total measurement time. The measurements will be displayed on the graph each time a new measurement is captured. The operator can select which temperature sensor will be displayed on the graph from the selector on the top right of the graph.

Report Generation Tab

The data in this tab are also entered manually according to the requirements of the customer. After filling in the necessary data, the report about the entire DC test can be generated by clicking the Generate Report button (Figure 6.42).

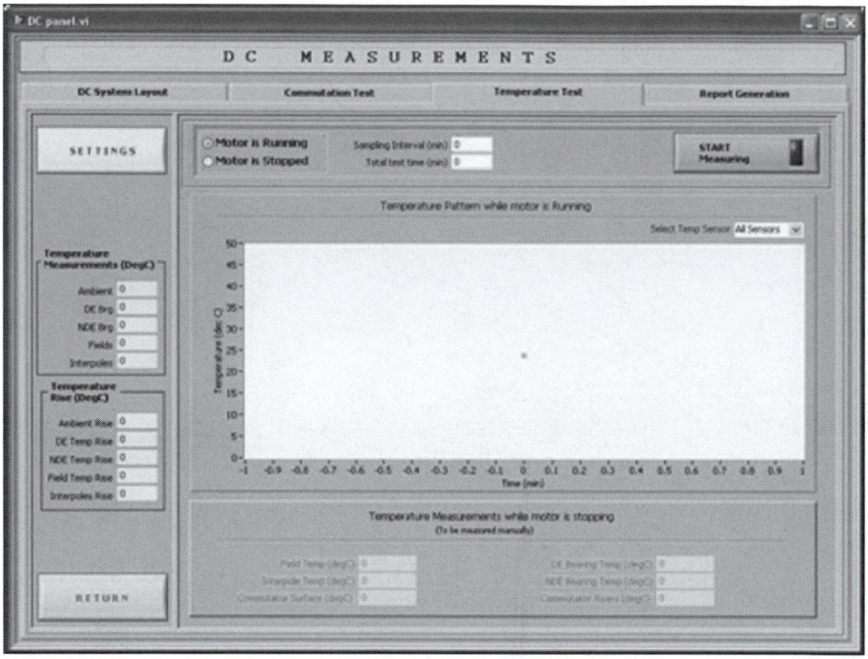

FIGURE 6.41   Front panel of the Temperature Test Tab.

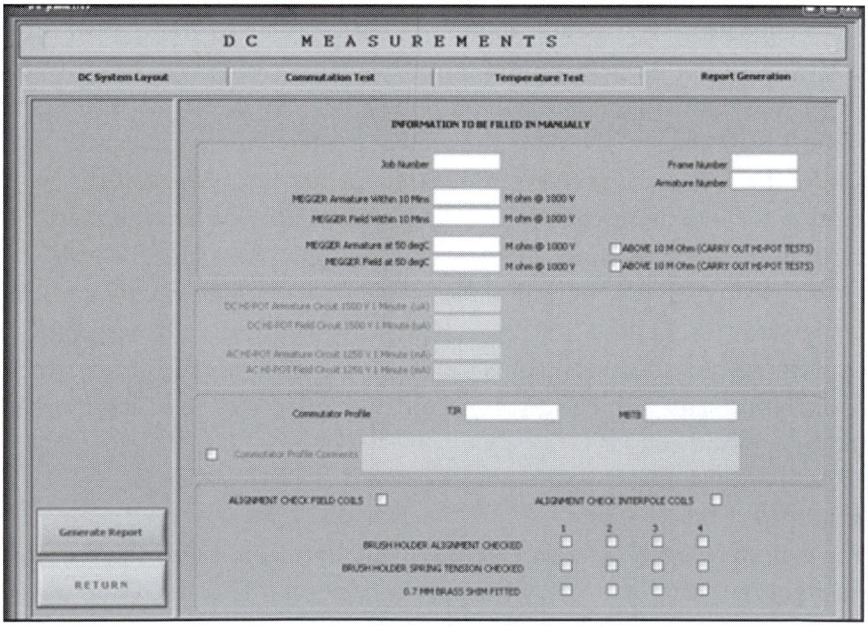

FIGURE 6.42   Front panel of the Report Generation Tab.

RESOURCES

A laptop computer with suitable DAQ card at a minimum sampling rate of 200 kHZ

Voltage and current sensors with suitable ratings to match the ratings of the electrical machine under tests (refer to Table 6.1 for the desirable sensor specifications)

REFERENCES

Bakhri, S. (2008). Investigation and Development of a Real-Time On-Site Condition Monitoring System for Induction Motors, MSc. thesis. University of Adelaide, Australia.

Gargoom, A. (2006). Digital Signal Processing Techniques for Improving the Automatic Classification of Power Quality Events, Ph.D. thesis. University of Adelaide, Australia.

CHAPTER 7

Recursive Computation of Discrete Wavelet Transform

Nasser Kehtarnavaz, Vanishree Gopalakrishna, and Philipos Loizou

CONTENTS

REAL-WORLD INTEGRATION

Wavelet transform is increasingly being used in place of short-time Fourier transform in various signal processing applications such as time-frequency analysis, data compression, denoising, and signal classification. While short-time Fourier transform deploys a fixed window length for all frequencies, wavelet transform deploys varying window lengths for different frequencies where lower frequencies are captured via larger window lengths and higher frequencies are captured via smaller window lengths.

Many signal processing applications require wavelet transform to be computed in real-time over a moving window. Due to the computational efficiency of discrete wavelet transform (DWT), in practice, this wavelet transform is often used in place of continuous wavelet transform. This chapter presents a recursive way of computing DWT over a moving window, which is shown to be computationally more efficient than the conventional nonrecursive approach.

INTRODUCTION AND BACKGROUND

Let us first provide a brief overview of DWT. Consider an input signal $x(n)$ of length N. DWT of $x(n)$ is defined in terms of so-called approximation coefficients $A(j_0, p)$ and detail coefficients $D(j, p)$ as follows:

$$A(j_0, p) = \frac{1}{N} \sum_{n=0}^{N-1} x(n) \varphi_{j_0, p}(n) \tag{7.1}$$

$$D(j, p) = \frac{1}{N} \sum_{n=0}^{N-1} x(n) \Upsilon_{j, p}(n) \tag{7.2}$$

where $j \geq j_0, j = 0, 1, \ldots, J-1, p = 0, 1, 2, \ldots, 2^j - 1$, and J denotes the number of decomposition levels such that $2^J \leq N, \varphi_{j_0, p}(n), \Upsilon_{j, p}(n)$ represent appropriate basis functions that are obtained by

$$\varphi_{j, p}(n) = 2^{j/2} \varphi(2^j n - p) \tag{7.3}$$

$$\Upsilon_{j, p}(n) = 2^{j/2} \Upsilon(2^j n - p) \tag{7.4}$$

based on a scaling function $\varphi(n)$ and a wavelet function $\Upsilon(n)$.

The approximation and detail coefficients can be used to reconstruct the signal $x(n)$ by using the Inverse DWT transform given by

$$x(n) = \frac{1}{N} \sum_{p=0}^{2^j - 1} A(j_0, p) \varphi_{j_0, p}(n) + \frac{1}{N} \sum_{j=j_0}^{J} \sum_{p=0}^{2^j - 1} D(j, p) \Upsilon_{j, p}(n) \tag{7.5}$$

As it is well established, DWT can be computed efficiently by using a filterbank structure (Mallat 1998; see Figure 7.1). This filterbank structure consists of a pair of decomposition (analysis) low pass and high pass filters. The detail coefficients $d1(n)$s are obtained by filtering $x(n)$ through a high pass filter $g(n)$ and then by downsampling the filtered output by a factor of 2. Similarly, the approximation coefficients $a1(n)$s are obtained by passing the input signal $x(n)$ through a low pass filter $h(n)$ and then by downsampling the filtered output by a factor of 2. The unit sample response of the low pass and high pass filters, $h(n)$ and $g(n)$, are known depending on the wavelet type or basis functions used. Sample magnitude responses of the high pass and low pass decomposition filters are shown in Figure 7.2.

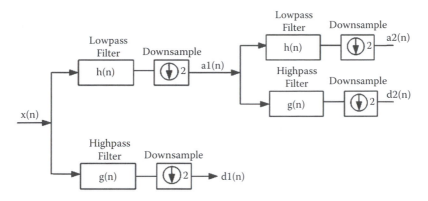

FIGURE 7.1   Filterbank structure for efficient computation of DWT.

The process of filtering and downsampling is continued until a specified frequency resolution is reached.

Let the highest frequency contained in the input signal $x(n)$ be ω. Then, the required minimum input sampling frequency is 2ω as per the Nyquist theorem. The highest-frequency content of the filtered signal after low pass filtering is $\omega/2$. Since the required minimum sampling frequency is ω, the filtered signal is downsampled by a factor of 2. The approximation coefficients correspond to the signal with the frequency range from 0 to $\omega/2$ and the detail coefficients to the frequency range from $\omega/2$ to ω. Downsampling by a factor of 2 reduces the time resolution and at the same time increases the frequency resolution by a factor of 2.

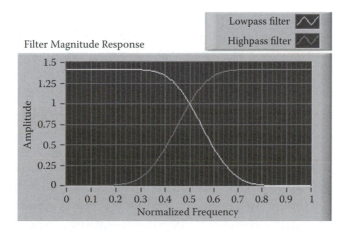

FIGURE 7.2   Magnitude responses of high pass and low pass decomposition filters for wavelet type "Symmlet 8."

The input signal can be reconstructed using the inverse wavelet transform via a filterbank similar to decomposition. The approximation and detail coefficients at each stage are upsampled by a factor of 2 by filling zeros in between consecutive samples and then by passing them through a reconstruction (synthesis) low pass $\tilde{h}(n)$ and a high pass $\tilde{g}(n)$ filter, respectively. The outputs from the high pass and low pass filters are summed up using the same number of stages used during the decomposition.

As discussed in (Strang and Nguyen 1996), Equations (7.6) and (7.7) give the conditions for exact reconstruction: (7.6) ensures there is no aliasing, and (7.7) ensures no distortion:

$$\tilde{H}(z)H(-z)+\tilde{G}(z)G(-z)=0 \tag{7.6}$$

$$\tilde{H}(z)H(z)+\tilde{G}(z)G(z)=2z^{-l} \tag{7.7}$$

For streaming signals, similar to short time Fourier transform, a moving window or frame is deployed over which DWT coefficients are computed. The computation of DWT coefficients over one window or frame is normally done independently of another frame. For real-time applications, this approach may limit a desired real-time processing rate. In this chapter, we present a method to compute the DWT coefficients of a frame recursively based on the previously computed coefficients of a previous frame. Such a recursive computation method is shown to involve fewer operations and is thus faster than the conventional nonrecursive computation method that is implemented in existing software packages.

The same recursive method can be applied to a binary tree wavelet packet transform (WPT; see Figure 7.3), where the decomposition at the second stage and beyond is also applied to the detail coefficients. Due to downsampling at each stage, the number of samples available at the output depends on the number of decomposition stages. Consider L_{stage} number of decomposition stages. For an input frame with N samples, the output consists of $2^{L_{stage}}$ frames of $N/2^{L_{stage}}$ samples. To obtain a higher output rate, one can have the input frame windowed and the window shifted such that there is an overlap between the shifted or moving windows, a concept commonly used in performing overlapping window short-time Fourier transform. By varying the amount of overlap, the output rate can be altered, the least being $N/2^{L_{stage}}$ for nonoverlapping windows and N when the window is shifted by one sample at a time.

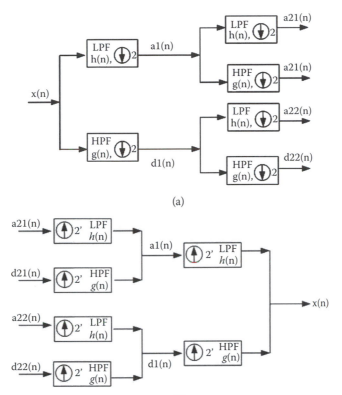

FIGURE 7.3 (a) Two-stage WPT decomposition or analysis filterbank. (b) Two-stage WPT reconstruction or synthesis filterbank.

An application where the overlapping window wavelet packet transform is used appears in Gopalakrishna, Kehtarnavaz, and Loizou (2010). In this application, a PDA platform was used to compute the wavelet packet transform in real time for cochlear implant research studies. It was shown that the conventional nonrecursive computation of the wavelet packet transform did not meet the required real-time constraint, whereas the recursive computation method managed to meet it.

CHALLENGE

When there is a large amount of overlap between the moving windows, the number of computations increases considerably compared with the non-overlapping moving window case. In real-time applications, this increase in the number of computations creates a bottleneck and poses an implementation challenge.

TABLE 7.1   Number of Multiplications Per Frame to Compute WPT

Method	Nonoverlapping Window	Overlapping Window
Number of multiplications per frame	$\left(\dfrac{N}{2^{L_{stage}}}\right)*\left[\displaystyle\sum_{j=1}^{L_{stage}}\left(\dfrac{2^{L_{stage}}}{2^{j}}\right)*l*(2^{j})\right]$	$\left(\dfrac{N}{2^{sh}}\right)*\left[\displaystyle\sum_{j=1}^{L_{stage}}\left(\dfrac{2^{L_{stage}}}{2^{j}}\right)*l*(2^{j})\right]$
$N=256$, $L_{stage}=3$, $2^{sh}=1$, $l=4$	≈ 3072	≈ 24576
$N=256$, $L_{stage}=7$, $2^{sh}=1$, $l=4$	≈ 7168	≈ 917504

Table 7.1 gives the increase in the number of multiplications for the nonoverlapping and overlapping cases where

N = number of samples per frame
L_{stage} = number of decomposition stages ($2^{L_{stage}} \leq N$),
2^{sh} = number of sample shifts per window ($\leq 2^{L_{stage}}$),
l = filter unit sample response length

SOLUTION: RECURSIVE COMPUTATION OF DWT AND WPT

Since the number of computations required for the overlapping window DWT becomes considerably higher compared with the nonoverlapping window DWT, a recursive computation method is devised and discussed in Gopalakrishna et al. (2010) to lower the amount of computation. This method uses the data redundancy in overlapping windows by reusing some of the outputs that have already been computed in the previous window.

The procedure to compute WPT coefficients using the recursive method for the overlapping window case is explained with the following pseudo-code, which describes how the recursive updating of wavelet coefficients is performed when the moving window is shifted by one sample at a time.

Let $\psi_{j,n}^{p}(1:N_{j})$ be a window of N_{j} samples at the j-th decomposition stage with branch number p at the time instant n. At the j-th stage, there are a total of 2^{j} branches. An input frame of N_{0} samples passed through a two-stage wavelet packet decomposition yields the outputs at four branches with N/4 samples, as shown in Figure 7.4.

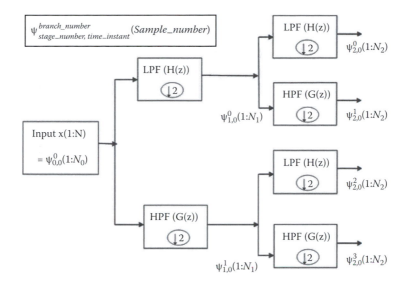

FIGURE 7.4   Two-stage WPT decomposition for a frame of samples.

Updating of the wavelet coefficients for a one sample shift of the moving window is achieved as follows:

```
for stage_number j = 1 to L_stage
        for branch_number p = 1 to 2^j
                update circular buffer pointers such that
                ψ^P_{j,n}(1:N_j − 1) = ψ^P_{j, n−2^j}(2:N_j)
                compute last sample of window ψ^P_{j,n}(N_j)
        end
end
```

The following equations give the updates for the $(j + 1)$-th stage at branches $2p$ and $2p + 1$:

$$\psi^{2p}_{j+1,n}(1:N_{j+1}-1)=\psi^{2p}_{j+1,n-2^{j+1}}(2:N_{j+1}) \tag{7.8}$$

$$\psi^{2p+1}_{j+1,n}(1:N_{j+1}-1)=\psi^{2p+1}_{j+1,n-2^{j+1}}(2:N_{j+1}) \tag{7.9}$$

$$\psi^{2p}_{j+1,n}(N_{j+1})=\sum_{k=0}^{l-1}\psi^{p}_{j,n}(N_j-k)*h(k) \tag{7.10}$$

$$\psi_{j+1,n}^{2p+1}(N_{j+1}) = \sum_{k=1}^{l-1} \psi_{j,n}^{p}(N_j - k) * g(k) \tag{7.11}$$

$$\forall\; j = 0,1,\dots, L_{stage} - 1, \quad p = 0,1\dots 2^j - 1, \quad n = 0,1,\dots,$$

and l is the decomposition filter length that depends on the wavelet basis and the scale chosen.

For a window containing N samples, the number of samples at the j-th stage output is $N_j = N/2^j$. It is easy to see that when the window is shifted by one sample, the corresponding windows in all the stages also get shifted by one sample. The time instance at which the window is used to perform the update depends on the stage. For example, for stage 1 the window used to perform the update is the window preceding the previous window. In general, at any given stage the window used to perform the update is the window that occurred $2^{stage\ number}$ instances before. This is illustrated in Figure 7.5 for a two-stage decomposition.

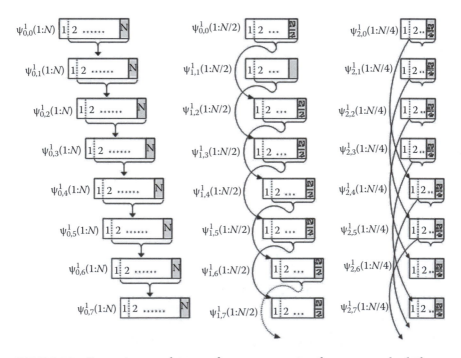

FIGURE 7.5 Recursive wavelet transform computation for one sample shift per window.

In general, for 2^{sh} sample shifts with $sh = 1,2,\ldots$, the updating procedure is given by

$$\psi_{j+1,n}^{2p}(1:N_{j+1}-1) = \psi_{j+1,n-2^{j+1-sh}}^{2p}(2:N_{j+1}), \tag{7.12}$$

$$\psi_{j+1,n}^{2p+1}(1:N_{j+1}-1) = \psi_{j+1,n-2^{j+1-sh}}^{2p+1}(2:N_{j+1}) \tag{7.13}$$

$$\psi_{j+1,n}^{2p}(N_{j+1}) = \sum_{k=0}^{l-1} \psi_{j,n}^{p}(N_j - k) * h(k) \tag{7.14}$$

$$\psi_{j+1,n}^{2p+1}(N_{j+1}) = \sum_{k=0}^{l-1} \psi_{j,n}^{p}(N_j - k) * g(k) \tag{7.15}$$

$\forall\ j = sh+1,\ldots L_{stage} - 1, \quad p = 0,1,\ldots 2^j - 1, \quad n = 0,1,\ldots.$ and

$$\psi_{j+1,n}^{2p}\left(1:N_{j+1}-2^{sh-(j+1)}\right) = \psi_{j+1,n-1}^{2p}\left(2^{sh-(j+1)}:N_{j+1}\right) \tag{7.16}$$

$$\psi_{j+1,n}^{2p+1}\left(1:N_{j+1}-2^{sh-(j+1)}\right) = \psi_{j+1,n-1}^{2p+1}\left(2^{sh-(j+1)}:N_{j+1}\right) \tag{7.17}$$

$$\psi_{j+1,n}^{2p}(N_{j+1}-\delta) = \Sigma_{k=0}^{l-1}\psi_{j,n}^{p}(N_j - 2\delta - k) * h(k) \tag{7.18}$$

$$\psi_{j+1,n}^{2p+1}(N_{j+1}-\delta) = \sum_{k=0}^{l-1} \psi_{j,n}^{p}(N_j - 2\delta - k) * g(k) \tag{7.19}$$

$\forall\ j = 0,1\ldots sh \quad \text{and} \quad \delta = 0.1\ldots 2^{sh-(j+1)} - 1.$

For 2^{sh} sample shifts, the memory required at a stage can also be computed as follows:

$$\text{Number of filters at stage } j = 2^j$$

Number of windows to be kept in memory for a filter at any branch of the j-th stage

$$= \begin{cases} 1 & , \quad j \le sh \\ 2^{j-sh} & , \quad j > sh \end{cases}$$

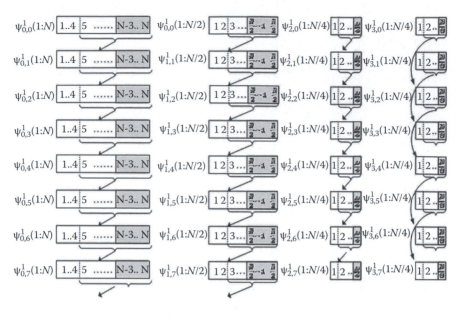

FIGURE 7.6   Recursive wavelet transform computation for four sample shifts per window.

Number of samples to be buffered in every window

$$
= \begin{cases} L-1+2^{sh-j}, & j \le sh \\ L, & j > sh \end{cases}
$$

This thus yields a total memory requirement of

$$
\approx \sum_{j=0}^{sh} (2^j * (l-1+2^{sh-j})) + \sum_{j=sh+1}^{L_{stage}-1} (2^j * l * 2^{j-sh}) \tag{7.20}
$$

The number of samples to be computed at every stage and the window used to do the update depends on the stage number as well as the number of sample shifts per window. As an example, the updating procedure for four sample shifts and a three-stage decomposition is illustrated in Figure 7.6.

The number of computations (multiplications) required for computing the wavelet packet transform using the nonrecursive and recursive methods is listed in Table 7.2 for the following specifications:

Frame length $= N(\ge 2^{L_{stage}})$

Number of sample shifts per window $= 2^{sh}(\le 2^{L_{stage}})$

TABLE 7.2   Number of Multiplications Required for the Nonrecursive and Recursive
Computation Methods

Method	Nonrecursive	Recursive
Number of multiplications	$\approx\left(\dfrac{N}{2^{sh}}\right)*\left[\displaystyle\sum_{j=1}^{L_{stage}}\left(\dfrac{2^{L_{stage}}}{2^j}\right)*l*(2^j)\right]$	$\approx\left(\dfrac{N}{2^{sh}}\right)*\left[\displaystyle\sum_{j=1}^{2^{sh}}2^{sh-j}*l*2^j+\displaystyle\sum_{j=sh+1}^{L_{stage}}l*2^j\right]$
$N=16,$ $L_{stage}=3,$ $2^{sh}=1, l=4$	≈ 1536	≈ 896
$N=256,$ $L_{stage}=8,$ $2^{sh}=1, l=4$	$\approx 2097152 \approx 2M$	$\approx 522240 \approx 522K$

Total number of windows $= N/2^{sh}$

Number of filters at any stage j, $F_j = 2^j$

Window size at decomposition stage j, $N_j = \dfrac{2^{L_{stage}}}{2^j}$

Number of multiplications at j-th stage to compute coefficients of a frame of length N_j is $l*N_j$

Total number of filters for L_{stage} decompositions $= \Sigma_{j=1}^{L_{stage}} 2^j = (2^{L_{stage}+1} - 2)$

Total number of multiplications for the conventional nonrecursive method

$$= (number\ of\ windows)\ *[(N_1 *l)*F_1 +(N_2 *l)*F_2 \ldots +(N_{L_{stage}} *l)*F_{L_{stage}}] \tag{7.21}$$

Total number of multiplications for the recursive method

$$= (number\ of\ windows) \left[\begin{array}{l}(2^{sh-1} *l)*F_1 +(2^{sh-2} *l)*F_2 \ldots +(2^{sh-sh} *l)*F_{sh} \\ +(1*l)*F_{sh+1} +\cdots+(1*l)*F_{L_{stage}}\end{array}\right] \tag{7.22}$$

LABVIEW IMPLEMENTATION

Many software tools including LabVIEW provide functions for computing DWT and WPT coefficients. In LabVIEW, wavelet coefficients can be found by using the function "WA Discrete Wavelet Transform VI" or "WA Undecimated Wavelet Transform VI" as part of the Advanced Signal

TABLE 7.3   Processing Time Required to Compute WPT Coefficients for 11 ms Frames

Method/Processing Time	Nonoverlapping Windows	Overlapping Windows ($2^{sh} = 1$)
$N = 256, L_{stage} = 3, l = 4$	1.9 *ms*	10.1 *ms*
$N = 256, L_{stage} = 7, l = 4$	2.2 *ms*	143.0 *ms*

Processing toolkit. The output rate provided by this function is fixed. In other words, one cannot choose a desired output rate depending on the amount of window overlap. It may be desired for the output rate to be even as high as the input rate or get adjusted depending on the amount of overlap between successive windows or frames. As discussed already, this creates a considerable computational burden if the computation of wavelet coefficients is done conventionally by treating moving windows and frames independently.

The increase in the number of operations gets translated into an increase in the processing time, which in turn limits the real-time throughput rate. As an example, Table 7.3 provides the processing time on a 3 GHz PC for computing the wavelet coefficients using the LabVIEW function "WA WP Decomposition VI" for the nonoverlapping and overlapping window cases. The input frame is of size 256, or 11.6 ms at a sampling rate of 22050 Hz. As shown in the table, the processing time increases considerably for the overlapping window case.

In our implementation, the functions to compute the DWT and WPT both nonrecursively (conventional) and recursively are written in C. These functions are built as Dynamic Link Libraries (DLLs) and are called using the "function call node" in LabVIEW, which can be found under "Functions >> Connectivity >> Libraries & Executables >> Call Library Function Node." The required parameters to call the functions are shown in Figure 7.7. Figure 7.7a shows how to call the wavelet tree data structure initialization, and Figures 7.7b and 7.7c show how to call the function for computing WPT and DWT coefficients as well as the inverse transform. Appropriate data types are chosen for each of the parameters passed. For example, all the single-byte arrays are passed as arrays of float with the parameter setting shown in Figure 7.8.

Table 7.4 gives the processing time required to compute WPT for the overlapping window case with one sample shift per window. The timing provided in the table was obtained using the conventional non-recursive and recursive C functions provided in the resource section. The frame size was chosen to be 256 samples, corresponding to 11.6 ms frames at the sampling frequency of 22050 Hz. The timings reported

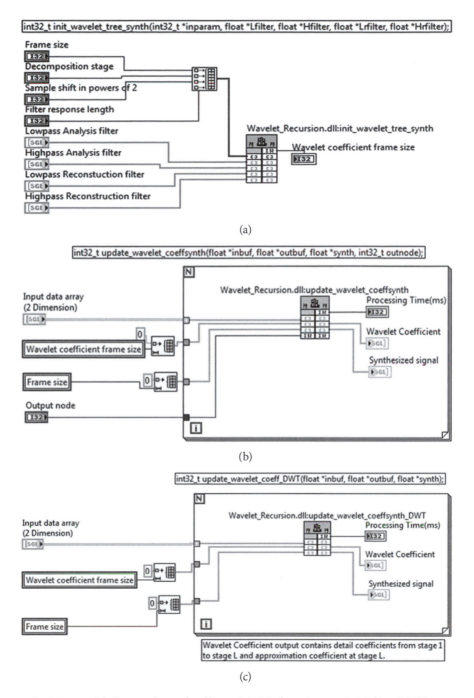

(a)

(b)

(c)

FIGURE 7.7 (a) Screenshot of calling C DLL function to initialize WPT tree structure. (b) Screenshot of calling C DLL function to compute WPT coefficients. (c) Screenshot of calling C DLL function to compute DWT coefficients.

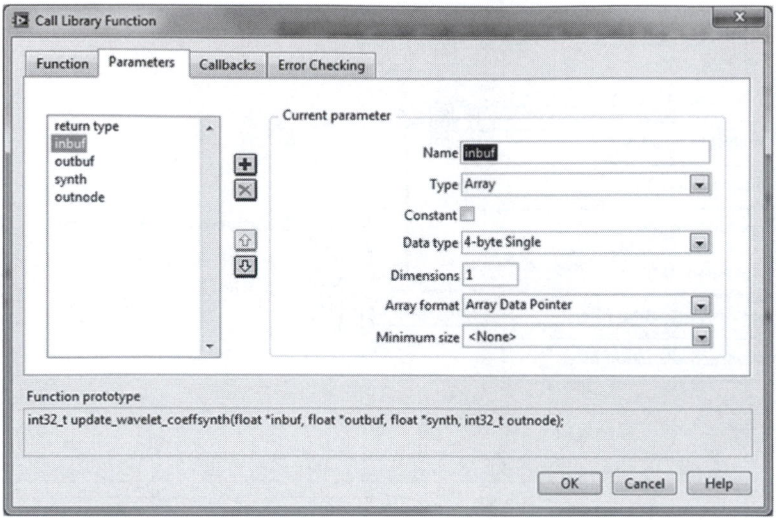

FIGURE 7.8   Settings for parameters passed to C DLL from LabVIEW.

in this table include the time required to make a call to the C function for every frame and the time to perform the decomposition as well as the reconstruction.

Two LabVIEW programs are provided at the link provided in the resource section: one reads a.wav file and plots the WPT/DWT coefficients for a specified output node; the other reads the input from a sound card. Both programs play the reconstructed signal. As shown in Figure 7.9, the parameters to choose before computing WPT/DWT include frame size to read from a.wav file, wavelet type, number of decomposition stages, number of sample shifts per window, output node for which the coefficients are to be displayed, and whether the nonrecursive or recursive method is to be used.

The first tab shows the WPT tree depending on the number of decomposition stages chosen and the unit sample and magnitude responses of the filters for a chosen wavelet type. A screenshot of the first tab showing the wavelet tree and the filters responses is shown in Figure 7.10.

TABLE 7.4   Processing Time Required to Compute WPT Coefficients Using the Nonrecursive and Recursive WPT Method

Method/Processing Time	Nonrecursive Method	Recursive Method
$N = 256$, $L_{stage} = 3$, $2^{sh} = 1$, $l = 4$	1.96 ms	0.97 ms
$N = 256$, $L_{stage} = 7$, $2^{sh} = 1$, $l = 4$	48.00 ms	10.08 ms

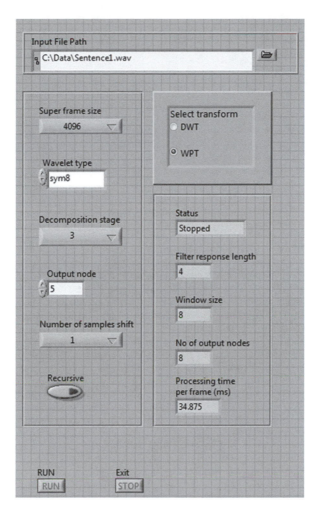

FIGURE 7.9 Screenshot of parameters to select for computing wavelet coefficients.

In the output tab, as shown in Figure 7.11, the input signal, the reconstructed or synthesized signal and the WPT coefficients of the chosen output node are plotted. The processing time taken to compute wavelet coefficients for a frame is also displayed. In general, one can see that the processing time using the recursive wavelet computation method is approximately one-half that of the nonrecursive or conventional method when the number of decomposition stages is 3 and is reduced to one-fifth when the number of decomposition stages is increased to 7.

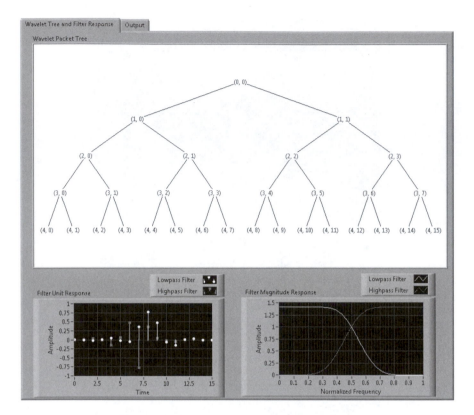

FIGURE 7.10 Screenshot showing the wavelet packet tree chosen and the unit and magnitude responses of the decomposition filters.

RESOURCES

The C functions written for the nonrecursive and recursive computation of WPT and DWT coefficients can be downloaded from www.utdallas.edu/~kehtar/WaveletRecursive/Ccodes.

The LabVIEW programs showing how to call these functions can also be downloaded from www.utdallas.edu/~kehtar/WaveletRecursive/LabVIEWcodes

In addition, two example programs, one using a.wav file and another performing the real-time on-the-fly wavelet computation of an input speech signal from a sound card, are provided at the above website.

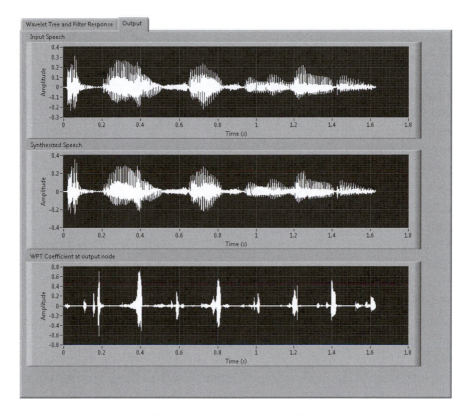

FIGURE 7.11 Output tab showing a speech signal, the reconstructed speech signal, and the WPT coefficient of a chosen output node.

REFERENCES

V. Gopalakrishna, N. Kehtarnavaz, and P. C. Loizou. (2010). A Recursive Wavelet-based Strategy for Real-Time Cochlear Implant Speech Processing on PDA Platforms, *IEEE Transactions on Biomedical Engineering* (7) 2053–2063.

S. Mallat. (1998). *A Wavelet Tour of Signal Processing.* Academic Press. San Diego.

G. Strang and T. Nguyen. (1996). *Wavelets and Filter Banks.* Wellesley-Cambridge Press.

CHAPTER 8

Solar Energy

Pedro Ponce, Brian MacCleery, Hiram Ponce, and Arturo Molina

CONTENTS

INTRODUCTION

The objective of this chapter is to introduce the design, development, and validation of solar tracking systems using LabVIEW software and other National Instruments devices. Photovoltaic cell modeling using LabVIEW is discussed using both simplified and more complex mathematical equations, intelligent modeling techniques, and characterization with LabVIEW. Solar tracking and aiming systems in LabVIEW are also presented as PID control techniques and genetic algorithms. Finally, solar tracking systems analysis is discussed using SolidWorks and LabVIEW, taking into account mechanical modeling motor design.

CHALLENGE

Finding sustainable ways to harvest and manage Earth's natural resources is fundamental to the economic development, national security, and environmental protection of the world. In these terms, renewable energy sources derived primarily from the sun's energy in the form of wind, hydro, and solar power are both the most ancient and the most modern forms of energy used by humans. Today, a new industrial revolution is under way to lower the cost and increase the production scale of these technologies to meet the world's ever-growing demand for energy.

Renewable energy is growing more important for a variety of reasons. While the bulk of energy production today comes from the burning of fossil fuels, the consequences and implications of its large-scale uses and the disadvantages with regard to human health and the environment are becoming well known. Furthermore, the nations of the world are increasingly seeking energy independence for national security reasons, since the economy and modern civilization itself require abundant access to energy.

This chapter focuses on solar energy in the form of radiant light from the sun and explores associated technologies and closely examines the advantages of solar tracking systems, an important technology for solar energy.

RESOURCES

The following resources were used during the development of the sun tracking system:

1. NI LabVIEW software

2. Intelligent Control Toolkit for LabVIEW

3. NI CompactRIO hardware

4. Dassault Systèmes SolidWorks software

5. Silicon photovoltaic cell

6. Stepper motors

7. Altitude-azimuth axis solar tracking mechanical system

BACKGROUND: SOLAR TRACKING SYSTEMS

Sun charts are tools developed to aid in the setting up of solar installations and determine during which times of the year a solar installation may become shaded. Sun charts are also useful for understanding the sun's path across the sky at different times of the year and the importance of sun tracking systems. Figure 8.1 shows a sun chart for Austin, Texas (latitude 30.4° in the Northern Hemisphere), as calculated and charted using the LabVIEW graphical programming language. The bottom trace, drawn on

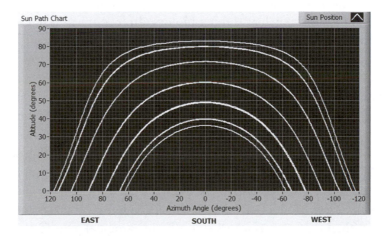

FIGURE 8.1   Sun chart for Austin, Texas (latitude 30.4°N), showing the path of the sun across the sky from December (bottom trace) to June (top trace).

the December 21 winter solstice, shows that the sun hangs low in the sky and follows a relatively circular path during the short days of the winter months. On the other hand, the June 21 summer solstice represents the longest day of the year in which the sun rises quickly to a nearly vertical elevation (altitude angle) and then traverses east to west (azimuth angle) across the sky before setting in the west. From July to November, the traces overlap other months as the seasons approach the winter solstice again.

Note that the average altitude of the sun during the brightest times of the year is equal to 90 degrees (directly vertical) minus the latitude (30.4°), in this case 59.6°. This would be the optimal altitude angle for a fixed tilt installation in which no seasonal adjustments are made.

Solar tracking systems are mechanical devices capable of orienting themselves relative to the sun's position as it moves across the sky. These systems can guide devices such as flat solar modules or arrays so they remain near perpendicular to the sunlight by tracking the sun from sunrise to sunset. In the case of a heliostat mirror tracking system, the goal is to keep the sunlight focused on a thermal collector target as the sun traverses the sky. In the case of concentrated photovoltaic (CPV) arrays, which use Fresnel lenses to concentrate the sunlight onto a small solar cell, extremely precise tracking accuracy is required. Due to the high magnification levels, the sunlight must be precisely focused on the small PV target.

Typical solar trackers are structurally formed by metals such as aluminum and stainless or painted steel. The structure must be able to withstand climatic changes, mainly rain and wind. The photovoltaic cells are attached to the top of the tracker. The motion of all structures is driven by motors, such as stepper motors, direct current (DC) motors, or position-controlled induction motors. Each motor has advantages for different applications, depending on the accuracy, load, speed, and pricing the system requires.

Advantages of solar trackers include their ability to enhance photovoltaic cell efficiency by 20 to 40%, depending on geographic location. The highest gain from tracking is for desert areas like the U.S. Mojave desert, where the sunlight is very direct rather than diffuse. In a location such as central Texas, for example, a single-axis tracker will increase annual power output by 26% compared with a fixed-angle system, whereas a dual-axis tracker increases production by 32%. Thus, single-axis trackers are more common than dual-axis trackers since the incremental gain for two-axis tracking is smaller than that from adding a single axis. Figure 8.2 shows two-axis tracking panels and south-facing PV panels with different tilts.

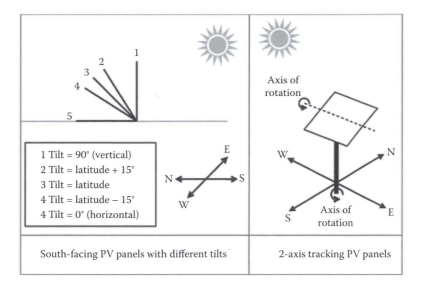

FIGURE 8.2 Two-axis tracking panels and south-facing PV panels. (From www. nrcan-rncan.gc.ca.)

Design considerations and challenges for sun tracking systems include the tracking accuracy, additional cost, energy consumption, ease of installation, ability to operate unattended, dust, corrosion and storm resistance, and the robustness and durability of the moving parts and actuators with respect to long-term operations and maintenance costs.

Depending on their mechanical and structural systems, solar trackers are classified as one of the following types.

FIXED AXIS

Fixed axis configurations involve a nonmotorized installation. For flat-panel photovoltaic arrays located in the Northern Hemisphere, the array is oriented due south with a manually adjusted tilt (elevation) angle. If minimal maintenance is the goal, the array orientation is set at the time of installation and never changed. For increased performance, seasonal adjustments are made to the elevation angle at specific dates during the year, typically on the March and September equinox. Seasonal adjustments are most important at the mid latitudes (closer to 45 degrees) in which the largest sun elevation differences between the winter and summer months occur. (These latitudes are also where automated sun tracking systems are most valuable.)

If no seasonal adjustments are made, the tilt angle (relative to a vertical orientation) is set equal to the latitude to optimize the average power output over the year. This provides good summer performance but reduced output in the spring and fall. This is a common configuration for grid-tied systems where low maintenance is required. However, lower annual production is achieved.

For off-grid systems using battery storage where seasonal adjustments are not desirable, the goal is to optimize performance during the winter months during which the solar irradiance is weakest and the longest periods of cloudy weather may occur. In this case, the tilt angle is set equal to the latitude plus 22.5 degrees to optimize winter performance.

TWO AXES

In this type of solar tracker, the surface of the photovoltaic cell is always perpendicular to the sunlight. This tracker mount is also becoming popular for large telescopes because of its structural simplicity and compact dimensions. One axis is a vertical pivot shaft or horizontal ring that allows the device to be swung to an east–west (azimuth) compass point. The second axis is a horizontal elevation pivot. By using combinations of two axes, any location in the upward hemisphere may be pointed (Figure 8.3). Such systems may be operated under computer control according to the calculated solar position or may use a tracking sensor to control motor drives that orient the panel toward the location of brightest solar irradiance.

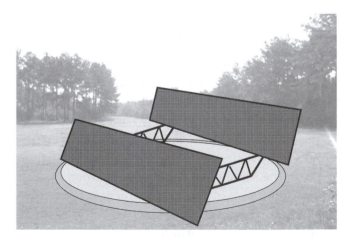

FIGURE 8.3   Two-axis solar tracking system.

POLAR AXIS

In a one-axis polar tracking configuration, the surface of the photovoltaic array is tilted vertically at an angle equal to the latitude of installation. The east–west rotation is adjusted so that the normal to the surface at all times coincides with the terrestrial meridian containing the sun. Polar trackers are often used in grid-tied configurations with time-of-use metering because performance is maximized during the typical time of peak electricity demand in the afternoon. Figure 8.4 shows an example of a polar-axis solar tracking system.

ALTITUDE–AZIMUTH AXIS

In this case, the photovoltaic cell surface rotates on a vertical axis. The angle of the surface is constant and equal to the latitude. The rotation is adjusted so that the normal to the surface at all times coincides with the local meridian containing the sun. One supporting axis is horizontal (called the altitude) and allows the photovoltaic cell to tilt up and down. The other axis (called the azimuth) and allows the cell surface to swing in a circle parallel to the ground to follow the sun east–west. A computerized motion control system keeps the photovoltaic cell surface oriented perpendicular to the sun. A typical altitude–azimuth axis solar tracking system is shown in Figure 8.5. This is the most popular type of solar tracker.

FIGURE 8.4 Polar-axis solar tracking system.

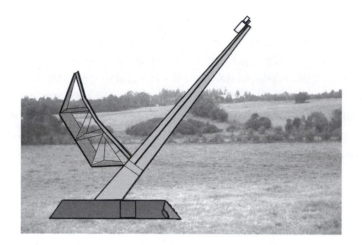

FIGURE 8.5   Altitude–azimuth axis solar tracking system.

HORIZONTAL AXIS

In this case, the surface of the solar cell rotates on a horizontal axis oriented north–south. The rotation is adjusted so that the normal to the surface at all times coincides with the terrestrial meridian containing the sun. Several manufacturers can deliver single-axis horizontal trackers, which may be oriented by either a passive or an active mechanism, depending on the manufacturer. Because these trackers do not tilt toward the equator, they are not especially effective during winter midday, but they add a substantial amount of productivity during the spring and summer seasons when the solar path is high in the sky. These devices are less effective at higher latitudes. The principal advantage is the simplicity of the mechanism. Figure 8.6 shows an example of this tracker system.

FIGURE 8.6   Horizontal-axis solar tracking system.

SOLAR TRACKING SYSTEMS USING LabVIEW

Solar tracking systems can be controlled via LabVIEW in a transparent and plug-and-play fashion. Conventional control such as proportional–integral–derivative (PID) or artificial intelligence methods can easily be used in a few steps with minimum embedded systems engineering expertise. Photovoltaic cell emulation, testing, and validation processes can also be conducted with LabVIEW to predict and measure the maximum power supply or efficiency.

PHOTOVOLTAIC CELL MODELING IN LabVIEW

The photovoltaic energy conversion in solar cells consists of two essential steps. First, the absorption of light generates an electron–hole pair. The electron and hole are then separated by the structure of the device—the electrons move to the negative terminal and the holes to the positive terminal, thus generating electrical power.

The solar cell can be considered as a two-terminal device that conducts like a diode in the dark and generates a photo voltage when charged by the sun. Usually, it is a thin slice of semiconductor material of around 100 cm² in area. The surface is treated to reflect as little visible light as possible and appears dark blue or black. A pattern of metal contacts is imprinted on the surface to make electrical contact.

Usually, when it is charged by the sun, this basic unit generates a DC photovoltage of 0.5 to 1 V and in short-circuit conditions a photocurrent of about 10 mA/cm². Although the current is reasonable, the voltage is too small for most applications. To produce useful DC voltages, the cells are connected in a series and encapsulated into modules. A module typically contains 28 to 36 cells in series to generate a DC output voltage of 12 V at standard illumination conditions. The 12 V modules can be used singly or connected in parallel and series in an array with a larger current and voltage output, according to the power demanded by the application.

Cells within a module are integrated with bypass and blocking diodes to avoid the complete loss of power that results if one cell in the series fails. Modules within the array are similarly protected. For almost all applications, the illumination varies too much for efficient operation all the time, and the photovoltaic generator must be integrated with a charge storage system (a battery) and components for power regulation. A model of the photovoltaic cell follows to illustrate obtaining a mathematical model for programming in LabVIEW.

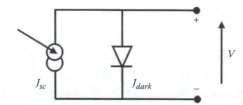

FIGURE 8.7   Equivalent circuit of the ideal solar cell.

SIMPLE EQUATION

An ideal solar cell can be represented by a current source connected in parallel with a rectifying diode, as shown in the equivalent circuit of Figure 8.7. The corresponding I-V characteristic is described by the Shockley solar cell equation

$$I = I_{PH} - I_0 \left(e^{\frac{qV}{k_B T}} - 1 \right) \tag{8.1}$$

where k_b is the Boltzmann constant ($1.3806503 \times 10^{-23} \frac{kgm^2}{s^2 K}$), T is the absolute temperature ($K°$), $q\,(>0)$ is the electron charge, V is the voltage at the terminal of the cell, I_{ph} is the photogenerated current (or short circuit current I_{sh}), and I_0 is the diode saturation current.

In terms of the current density

$$I = \int_s J \, ds$$

where s is the cross section area, and J is

$$J = J_{PH} - J \left(e^{\frac{qV}{k_B T}} - 1 \right)$$

the photogenerated current I_{ph} is closely related to the photon flux incident on the cell, and its dependence on the wavelength of light is frequently discussed in terms of the quantum efficiency or spectral response.

Figure 8.8 shows the I-V characteristics. In the ideal case, the short-circuit current I_{sh} is equal to the photogenerated current I_{ph}, and the open-circuit voltage V_{OC} is given by

$$V_{OC} = \frac{k_B T}{q} \ln \left(1 + \frac{I_{ph}}{I_0} \right)$$

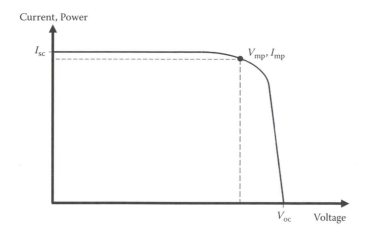

FIGURE 8.8 I-V characteristic of the solar cell and the maximum power generated.

The cell generates the maximum power P_{max} at a voltage V_m and current I_m, and it is convenient to define the fill factor FF by

$$FF = \frac{P_{max}}{I_{SC}V_{OC}} = \frac{I_m V_m}{I_{SC}V_{OC}}$$

The efficiency η of the cell is the power delivered at the operating point (P_{max}) as a fraction of the incident light power P_S

$$\eta = \frac{I_m V_m}{P_S}$$

Efficiency is related to I_{sh} and V_{OC} using FF. These four quantities—I_{ph}, V_{OC}, FF, and η—are the key performance characteristics of a solar cell

$$\eta = \frac{I_{SC}V_{OC}FF}{P_S}$$

COMPLEX EQUATION

In practice, real solar cell power is dissipated through the resistance of the contacts and through leakage currents around the sides of the device. These effects are electrically equal to two parasitic resistances in series (R_S) and in parallel (R_{SH}) with the cell. The equivalent circuit is shown in Figure 8.9.

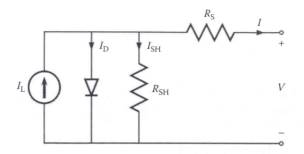

FIGURE 8.9   Equivalent circuit of a solar cell with series and shunt resistance.

The series resistance arises from the resistance of the cell material to current flow, particularly through the front surface to the contacts, and from resistive contacts. Series resistance is a particular problem at high-current densities, for instance, under concentrated light. The parallel or shunt resistance arises from the leakage of current through the cell, around the edges of the device, and between contacts of different polarity. It is a problem in poorly rectifying devices. Series and parallel resistances reduce the fill factor as shown in Figures 8.10a and 8.10b.

In an efficient cell, the R_S is small, and the R_{SH} is as large as possible. When parasitic resistances are included, the complex model equation is written as in (8.2), with n typically between 1 and 2

$$I = I_{PH} - I_0 \left(e^{\frac{q(V+IR_S)}{nk_BT}} - 1 \right) - \frac{V + IR_S}{R_{SH}} \tag{8.2}$$

MATHEMATICAL PHOTOVOLTAIC CELL EQUATIONS IN LabVIEW

Using Equations (8.1) and (8.2), a mathematical model of the photovoltaic cell can be programmed in LabVIEW. Figure 8.11 shows the control panel of the VI in which several parameters need to be introduced, such as solar incidence, solar cell materials, resistances, and the simple or complex information requested. Taking this model as the base model, nonlinear effects could be added to have a closer approximation of the photovoltaic cell.

Figure 8.12 shows the block diagram for the ideal or simple case model (from 8.1), and Figure 8.13 shows the block diagram for the complex or real case model (from 8.2). Differing surface areas could be used for testing the solar cell performance. Additionally, Figure 8.14 shows the parameters

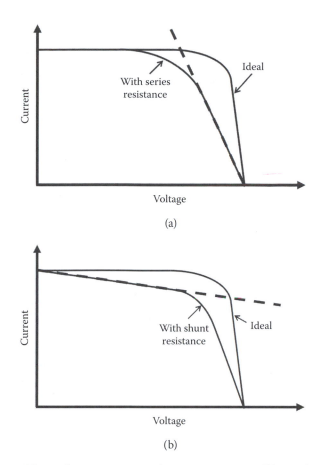

FIGURE 8.10 (a) I-V characteristics with series resistance. (b) I-V characteristics with shunt resistance.

(efficiency, V_{OC}, I_{SC}, FF, and cell area) that are returned by selecting the materials of the photovoltaic cell.

INTELLIGENT PHOTOVOLTAIC CELL MODELING IN LabVIEW

Other techniques for modeling are being proved. In this case, artificial intelligence methods were used for understanding the behavior of the photovoltaic cell. In particular, the Intelligent Control Toolkit for LabVIEW (ICTL) developed by ITESM University in Mexico City was used.

The ICTL has classical and novel algorithms such as artificial neural networks, fuzzy logic, neuro-fuzzy systems, genetic algorithms, genetic programming, and predictive algorithms in terms of artificial intelligence methods. The toolkit combines LabVIEW ease of use with some powerful

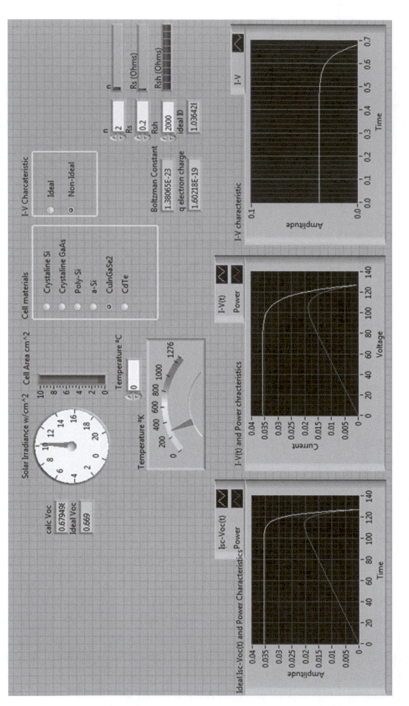

FIGURE 8.11 Control panel of the solar VI.

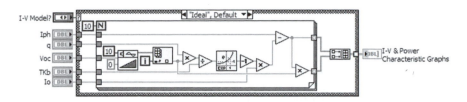

FIGURE 8.12   Simple ("ideal") photovoltaic model equation in LabVIEW.

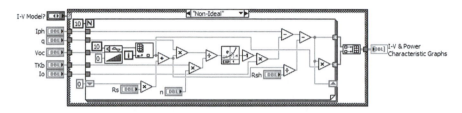

FIGURE 8.13   Complex ("realistic") photovoltaic model equation in LabVIEW.

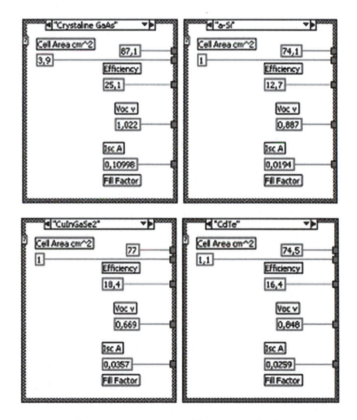

FIGURE 8.14   Several parameters returned by material selection.

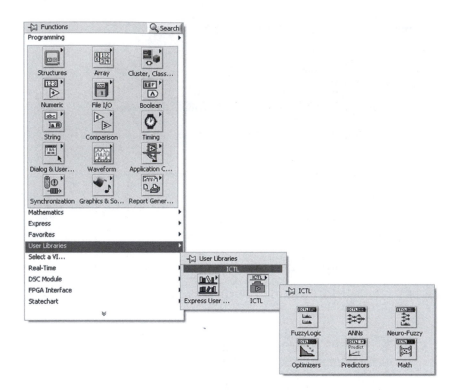

FIGURE 8.15   Location of the Intelligent Control Toolkit for LabVIEW.

intelligent control techniques. Even though it was originally created to be applied in the field of intelligent control, the ICTL has a wider range of areas where it can be used, such as economics.

In LabVIEW, the ICTL is accessed as a third-party resource located under the User Libraries tab of the Functions Palette. This tab is divided into six sections: Fuzzy Logic, ANNs (artificial neural networks), Neuro-Fuzzy, Optimizers, Predictors, and Math (Figure 8.15).

The first group within the ICTL palette contains several VIs for designing fuzzy control systems; the second group encompasses a variety of artificial neural networks and their trainings. The Optimizers group features genetic algorithms and genetic programming as well as a few algorithms on search–space optimization and optimal clustering. The Predictors group contains algorithms dedicated for forecasting, and the Math group includes some VIs used for supporting the aforementioned techniques. For more information, please visit http://www.tribalengineering.com/technology/ictl-ictl.aspx.

A fuzzy Sugeno system was designed for this purpose (Figure 8.16 shows its front panel). First, two inputs were defined as the solar radiation

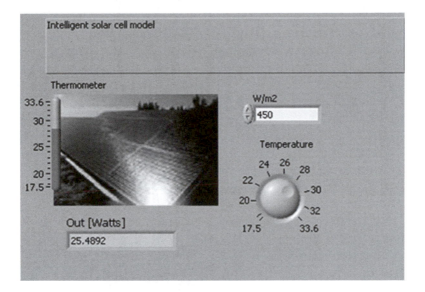

FIGURE 8.16   Control panel of the intelligent model of the photovoltaic cell in LabVIEW.

and the temperature at the surface of the cell. Then eight membership functions were assigned to each input with linguistic labels ranging from *nothing* through *several* to *too much*. Figure 8.17 shows these input membership functions. For instance, output functions of the Sugeno system are represented by singleton values according to the output power generated. A total of 64 rules were used to model the photovoltaic cell response. Figure 8.18 shows the block diagram of the fuzzy model. Some advantages of fuzzy systems are that linguistic categories may be used to understand the relationship between the inputs and outputs as a set of expert rules that need to be written in a mathematical formulation. Through these techniques, expert knowledge can be captured on these systems.

As an example of this fuzzy inference system, Figure 8.19 shows a photovoltaic cell model assuming a 25°C temperature at the surface of the cell and different solar radiation levels. In this graph, the fuzzy inference system model of a photovoltaic cell (labeled as "Fuzzy Model") is compared with measures of the power collected by the photovoltaic cell using different solar radiation levels (labeled as "Real Data"). As seen in graphs, the fuzzy inference system can reasonably approximate the real photovoltaic cell behavior.

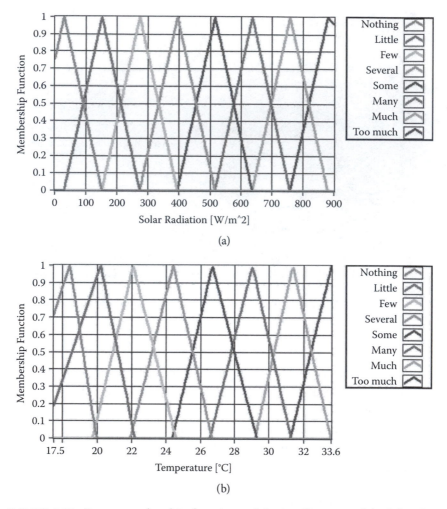

FIGURE 8.17 Input membership functions of the intelligent model of the photovoltaic cell.

PHOTOVOLTAIC CELL CHARACTERIZATION IN LabVIEW

LabVIEW offers another toolkit for analyzing and characterizing the relationship between the inputs of sunlight and temperature and the output power generated by the photovoltaic cell, called the "Toolkit for I-V Characterization of Photovoltaic Cells" (http://zone.ni.com/devzone/cda/epd/p/id/5918#1requirements).

Characterization is simple using this toolkit with any National Instruments data acquisition targets, especially source measurement units (SMUs), defined as power precision source instruments that provide

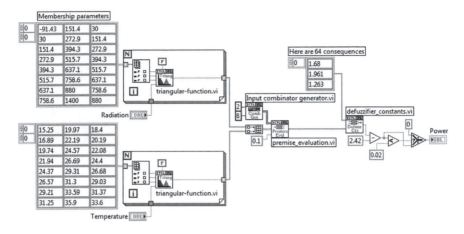

FIGURE 8.18 Block diagram of the photovoltaic cell fuzzy model in LabVIEW.

voltage and current sourcing. Figure 8.20 shows how this works and illustrates an instrumentation system used for report generation and analysis of the solar cell.

Figure 8.21 shows the control panel of the I-V characterization of a photovoltaic cell in which an SMU is used for acquiring the voltage and current values of the output photovoltaic cell. In the control panel, two inputs are required: (1) the area of the cell; and (2) the light intensity irradiance. Additionally, the I-V chart is plotted and parameters such as

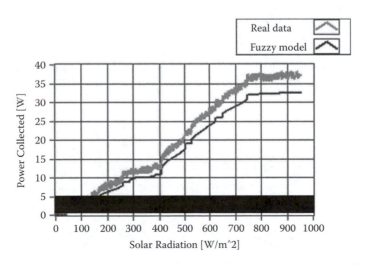

FIGURE 8.19 Result of the fuzzy model of the solar tracking system in LabVIEW.

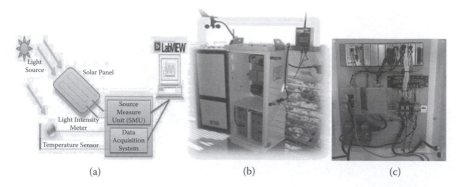

(a) (b) (c)

FIGURE 8.20 Characterization of the I-V photovoltaic cell using LabVIEW (a) and instrumentation systems using a NI CompactRIO (b and c).

maximum power computed. The voltage, current, performance parameters, and efficiency are also displayed. Figure 8.22 shows the block diagram of this VI.

Some information about the performance of this I-V characterization is summarized in Figure 8.23, taking into account advantages, drawbacks, and caveats for this practice.

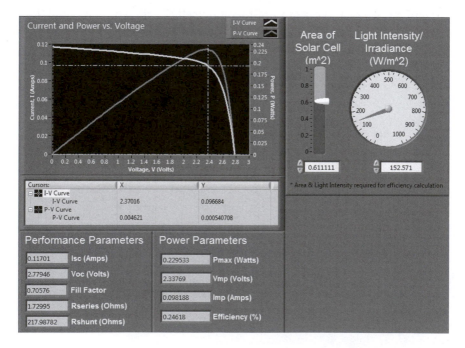

FIGURE 8.21 Front panel of the I-V characterization of a photovoltaic cell.

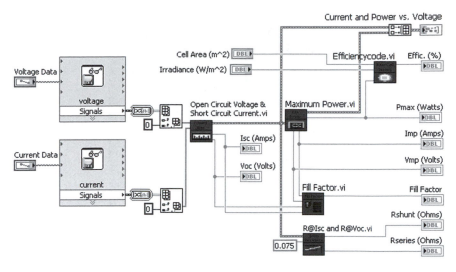

FIGURE 8.22   Block diagram of the I-V characterization of a photovoltaic cell.

SOLAR TRACKING SYSTEMS IN LᴀʙVIEW

At this point, solar cells are modeled with three different methods: (1) mathematical equations; (2) fuzzy inference systems; and (3) I-V characterization using LabVIEW. Then solar tracking systems need to incorporate control techniques for following the sunlight to obtain the maximum solar radiation and, consequently, the maximum power.

This section examines two kinds of controls. The first introduces PID control, whereas the second uses genetic algorithms to determine the optimal values at which a solar tracking system might be positioned over a day for collecting the most sunlight possible.

Advantages	Drawbacks	Caveats
Observation of performance before real operation	Power supply depends on the load connected to the photovoltaic cell	Necessity of previous characterization of the sensors
Flexibility of the system for any photovoltaic cell	Simulation does not incorporate climatic conditions such as clouds	Necessity of batteries
Simulation to prevent errors during operation		

FIGURE 8.23   Table of relevant information about I-V characterization of photovoltaic cells using LabVIEW.

FIGURE 8.24   Altitude-azimuth axis solar tracking demonstration system by National Instruments.

PID Control for Solar Tracking Systems Using LabVIEW

National Instruments developed an altitude–azimuth axis solar tracking system, as seen in Figure 8.24 (read more at **http://zone.ni.com/ devzone/cda/epd/p/id/6252**). A PID control refers to a proportional–integral–derivative control that aims to regulate a variable depending on a reference value.

National Instruments offers the LabVIEW Control Design and Simulation Module in which a PID control is already programmed. This VI needs the reference value, the gains, and the output range. The output of this function returns the correcting signal, which, in this case, is interpreted as the duty cycle of a pulse width modulation (PWM) that allows for moving any of the two motors for the altitude or the azimuth axes. Figure 8.25

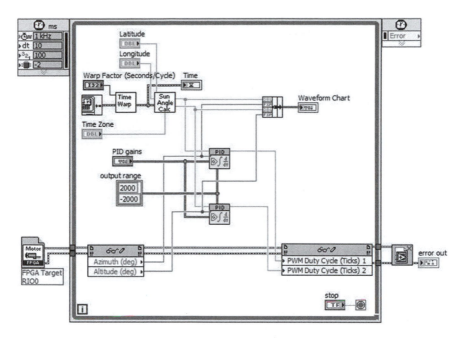

FIGURE 8.25   Block diagram of the PID control for a solar tracking system.

shows the block diagram of this PID control in which the reference is taken from the current sun position (measured via global positioning system, or GPS) and the duty cycle values are sent to the NI CompactRIO system. Figure 8.26 shows the control panel in which several parameters are displayed such as the voltage obtained by the photovoltaic cells.

Genetic Algorithms for a Solar Tracking System Using LabVIEW

The solar tracking system was optimized by genetic algorithms. The genetic algorithms find the optimal position configurations in which the photovoltaic cell collects the most sunlight possible while minimizing the motor power waste and steps involved in the process. For this purpose, an individual for genetic algorithms might be designed as $i = \{b_1, b_2, \ldots, b_{144}\}$, in which each bit b_n represents whether the photovoltaic cell moves to the ideal best position (predicted by the photovoltaic cell model and the current sun position) with the value **TRUE** or whether the photovoltaic cell stands by in the previous position (**FALSE**). Notice that each bit b_n represents the system configuration each minute, which is why there are 144 bits for the 144 minutes over a day. Thus, the best solution expected is a set of position configurations during a day with a minimum resolution of 10 minutes.

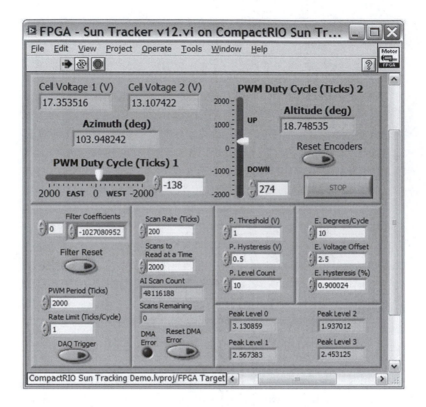

FIGURE 8.26   Control panel of the FPGA control for a solar tracking system.

Additionally, the fitness function may be developed to reach the goal of this optimization: maximizing the collection of solar radiation and minimizing the power used by motors. Thus, Figure 8.27 shows the block diagram of this fitness function.

The fitness function is created with the quadratic error average. For maximizing the power energy collected, a maximum upper limit (for example, 36 W is the maximum power supply by a given photovoltaic cell) has to be selected, such as 40 W. Then the power collected at the given time $P_{collected}^{t}$ is compared with this upper limit. The error is then powered by two, and the sum of these quadratic errors is divided by the maximum number of time values (144 in this case). Note that the maximum power collected per day is 3364 W for this example. Thus, in terms of optimization the error has to be decreased, and a better evaluation is needed. To determine this, the difference between the maximum power per day and the quadratic error is used. Then normalization is required. Second, to minimize the power

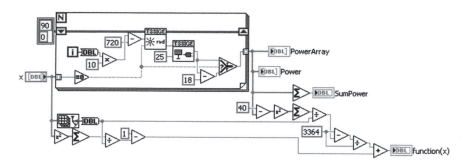

FIGURE 8.27 Fitness function used in the genetic algorithm for a solar tracking system.

used to drive motors, the sum of the quadratic error has to be minimized. Following the previous process, the normalized evaluation is found for the number of times that motors are switched on B_{action}^t. In mathematical terms, the fitness function is written as

$$f(x) = \left[1 - \frac{1}{(3364)(144)} \sum_{t=1}^{144} \left(40 - P_{collected}^t\right)^2\right] + \left[1 - \frac{1}{144} \sum_{t=1}^{144} \left(0 - B_{action}^t\right)^2\right]$$

In this way, the genetic algorithm can be run and can find the optimal values. Continuing with the example, five individuals was the population size with 0.95 of crossover probability and 0.01 of mutation probability. Figure 8.28 summarizes the set of tracking angle configurations and the time values for each configuration. Figure 8.29 presents the implementation of the optimization.

Analyzing the response of the optimal angles for altitude and azimuth axis angles, it can be seen that not all points are feasible. Thus, some considerations may be implemented and optimal values may be restructured, as seen in Figure 8.30.

REASONS FOR USING SOLAR TRACKING SYSTEMS

As discussed previously, solar tracking systems increase photovoltaic cell efficiency. These systems follow the sun's position to achieve the maximum sunlight that, ideally, is perpendicular to the photovoltaic cell surface. Efficiency increases to a maximum of 40%. This means that the remaining 60% of the energy is lost. However, this value is better than the approximately 80% energy lost when not using solar tracking systems.

Time [min]	Altitude [°]	Azimuth [°]	Time [min]	Altitude [°]	Azimuth [°]	Time [min]	Altitude [°]	Azimuth [°]
00:00:00	90	0	07:50:00	73	120	16:10:00	73	240
01:20:00	161	99	09:30:00	54	136	16:50:00	81	244
01:30:00	159	99	10:00:00	50	142	17:30:00	90	248
01:50:00	154	99	11:20:00	41	166	17:40:00	92	249
02:00:00	152	99	11:30:00	41	169	18:00:00	97	250
02:10:00	149	99	12:50:00	42	198	18:10:00	99	251
02:20:00	147	100	13:00:00	43	201	19:20:00	115	255
02:40:00	142	100	13:10:00	44	204	19:40:00	119	256
03:00:00	138	101	13:20:00	45	207	20:20:00	129	258
03:40:00	129	102	14:00:00	50	218	20:50:00	135	259
04:30:00	117	104	14:10:00	51	220	21:00:00	138	259
05:30:00	103	108	14:40:00	56	226	21:10:00	140	260
06:00:00	97	110	14:50:00	58	228	21:40:00	147	260
06:30:00	90	112	15:50:00	69	237	22:10:00	154	261
07:00:00	84	115	16:00:00	71	239	23:30:00	173	253

FIGURE 8.28   Results of aiming angles in the solar tracking system.

This is the main advantage of the solar tracker. It can follow the sun over the day. However, these trackers also involve several disadvantages, including the energy lost in driving motors. To overcome these challenges, a wide range of research must be conducted on solar tracking technologies.

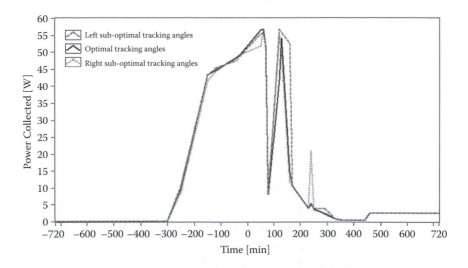

FIGURE 8.29   Response of the solar tracking system optimization.

Time [min]	Altitude [°]	Azimuth [°]		Time [min]	Altitude [°]	Azimuth [°]
00:00:00	90	0		14:00:00	50	218
09:30:00	54	136		14:10:00	51	220
10:00:00	50	142		14:40:00	56	226
11:20:00	41	166		14:50:00	58	228
11:30:00	41	169		15:50:00	69	237
12:50:00	42	198		16:00:00	71	239
13:00:00	43	201		16:10:00	73	240
13:10:00	44	204		16:50:00	81	244
13:20:00	45	207				

FIGURE 8.30   Final optimal angles of the solar tracking system.

This section described LabVIEW features to help explore the advantages of solar tracking systems. The next section turns to the mechanical structures involved in these systems. Computer analysis is introduced using SolidWorks and the embedded capabilities of LabVIEW for working with SolidWorks simulations of solar tracking systems.

SOLAR TRACKING SYSTEM ANALYSIS USING SolidWorks AND LabVIEW

The solar power industry is working to develop better results while reducing costs. In particular, mechanical structures and mechanical design use 3-D prototypes for testing and validating new approaches. However, making prototypes directly involves a high investment, and first prototypes do not guarantee the final solution. Thus, 3-D computer-aided design (CAD) programs such as SolidWorks are available.

Studies have shown that 72% of the industry's mechanical design applications involve 3-D CAD. Moreover, SolidWorks is one of the most used 3-D tools in industry. The LabVIEW platform is also popular. Thus, thanks to the NI SoftMotion application that links SolidWorks and LabVIEW, a new kind of powerful tool was created. This tool can reduce cost and time and optimize results that can be tested and validated with the control and simulation techniques in LabVIEW. This is useful in designing solar tracking systems. Notice that using the LabVIEW Control Design & Simulation Module photovoltaic cell modeling and mechanical structure modeling can also be integrated to implement the complex analysis that might be difficult to accomplish using other methods.

The next section introduces SolidWorks as a powerful tool for the 3-D design of mechanical structures for solar tracking systems. The end of the section presents the SolidWorks and LabVIEW interface for testing and validating the embedding solar tracking system.

SOLIDWORKS OVERVIEW

SolidWorks is 3-D CAD software that offers product data management, design validation, and CAD tools. Scientists and engineers use this platform to reduce design time and costs. Thus, SolidWorks is a powerful tool for these kinds of applications (Figure 8.31).

In addition, with SolidWorks, users can test products for errors before manufacturing, which helps prevent errors in the early steps of the design process and optimize designs for maximum performance. Mechanical CAD tools to test natural phenomena such as stresses, impact, heat, and airflow can be integrated into SolidWorks simulation applications to test prototypes.

Additionally, fluid flows can be tested as radiation heat transfer analysis, internal flow analysis, rotating reference frame analysis, transient flow analysis, and heat transfer analysis conduction and convection. Other SolidWorks applications include 3-D mechanical design, alternatives comparison, simulation embedding, mechanical simulation, and load cycles.

Mechanical Modeling in SolidWorks

This section describes a model assembly implemented in SolidWorks. A new project assembly must be started in SolidWorks. Different models of mechanisms for solar tracking systems can be made. For example, an altitude–azimuth axis, horizontal axis, or any other solar tracking system can be created. Specifically pay attention to the number of axes and type of motors selected. Axes and motors are the only entities that

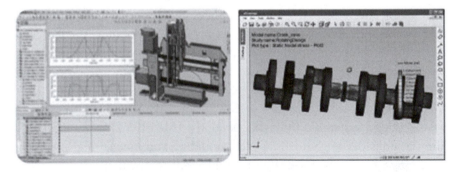

FIGURE 8.31 Example of a SolidWorks Assembly.

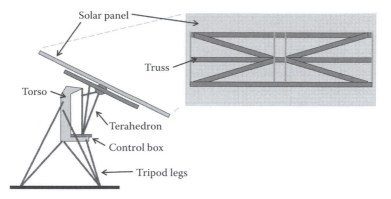

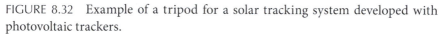

FIGURE 8.32   Example of a tripod for a solar tracking system developed with photovoltaic trackers.

this section examines. Note that previous knowledge of assembly designs on SolidWorks is required. Profiles, movements, natural phenomena, and constraints are part of the design process for solar tracking system mechanical structures on SolidWorks. Designing mechanical structures is not the only important feature. Constraints for analyzing stresses, impact, heat, and airflow onto the mechanical model are also important for the dimensional design on parts and for the selection of the motors, which are assessed by their torque and power.

Mechanical Structure

In some cases, a tripod type design is used to mount the photovoltaic array and tracking mechanicals. This structure can be made of aluminum, stainless steel, or any other material that prevents water corrosion, and it must be as robust as possible. Figure 8.32 shows the tripod of the solar tracking system. Basically, the system is composed of a tripod with helical pier anchors, the control box featuring the control signals to drive motors, a tetrahedron or arm to maintain the photovoltaic cell in the upper side of the tracker, and the truss to attach the solar array. Other mechanical structures differ in the arrangement of these parts based on specific needs or axis configuration.

STEPPER MOTORS

This example mechanical structure incorporates stepper motors, which are brushless, synchronous electric motors that can divide a full rotation into a large number of steps. This capability is due to the fact that a stepper motor does not need a feedback control mechanism to keep a position.

Another important advantage is that the largest of the step movements depends on the largest of the input voltage steps.

Stepper motor advantages also include a motor rotation angle that is proportional to the input pulse, full torque at standstill (if the windings are energized), accuracy of 3 to 5% of a step, and errors that are noncumulative from one step to the next. Additionally, these motors feature an excellent response to starting/stopping/reversing and to digital input pulses, providing open-loop control that makes the motor simpler and less expensive to control. In fact, it is possible to achieve very low-speed synchronous rotation with a load that is directly coupled to the shaft. However, stepper motors have several disadvantages. They lose torque when speed is increased and they lose precision if they do not have a control loop. They also need a stepper drive to move, exhibit rough performance at low speeds even when using microstepping, offer a limited size, and produce noise. The three common types of stepper motors are variable reluctance, permanent magnet, and hybrid. Figures 8.33a and 8.33b show these types of stepper motors.

The stepper motor principle of operation can be easily seen in the variable reluctance stepper motor, but it is common in all types of stepper motors. When a single winding or phase is energized, the motor generates a torque to align the rotor teeth with the teeth of the energized phase. The torque generated by current in a single phase is defined as

$$T_e = -k_T i_a \sin(N_r \theta)$$

where T_e is the generated torque, k_T is the torque constant, i_a is the current in phase, N_r is the number of electrical cycles per mechanical revolution, and θ is the mechanical rotor position.

Mathematical Models of Stepper Motors

In this example, two model stepper motors have been developed for variable reluctance: permanent magnet, and hybrid stepper motors. Initially, the equivalent circuit for one phase of a variable reluctance stepper motor is assumed. It is shown in Figure 8.34.

Consider the following: Ferromagnetic material does not saturate, and the inductance for each phase varies as a sinusoidal form around the circumference of the air gap. In other words, at A-phase, $L_A(\theta) = L_0 + L_1 \cos(N_r \theta)$, where L_0 is the average inductance, L_1 is the maximum inductance variation, and N_r is the rotor teeth number. Note that at the reference position ($\theta = 0$), the rotor tooth is fully aligned with an A-axis pole so that the

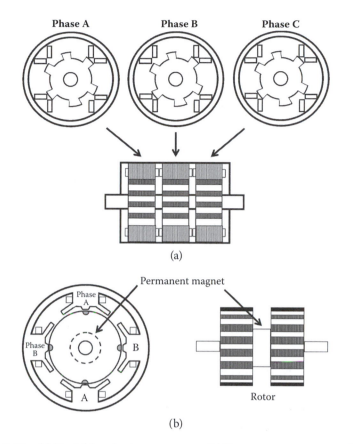

FIGURE 8.33 (a) Variable reluctance stepper. (b) Permanent magnet, and hybrid stepper.

A-phase winding inductance is at its maximum. The saturation problem is addressed later.

According to the last paragraph, the terminal voltage for phase A can be found using Faraday's law as

$$V_a = R_a i_a + \frac{d\lambda_a}{dt}$$

where V_A is the terminal voltage, R_A is the winding resistance, i_A is the winding current, and λ_A is the phase flux linkages. Because $\lambda_A = L_A i_A$

$$\frac{di_A}{dt} = \frac{1}{L_A}V_A - \frac{R_A i_A}{L_A} + \frac{L_1 N_r}{L_A} i_A w \sin(N_r \theta)$$

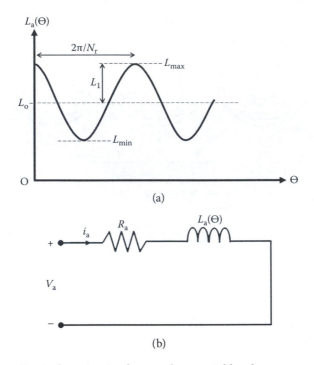

FIGURE 8.34   Equivalent circuit of a one-phase variable reluctance stepper motor.

Inductances for the other phases, however, need to be shifted into position. For a three-phase motor, the inductances are

$$L_B(\theta) = L_0 + L_1 \cos\left(N_r\theta - \frac{\pi}{3}\right)$$

$$L_C(\theta) = L_0 + L_1 \cos\left(N_r\theta = \frac{2\pi}{3}\right)$$

Optionally, for a four-phase motor the inductances are

$$L_B(\theta) = L_0 + L_1 \cos\left(N_r\theta - \frac{\pi}{2}\right)$$

$$L_C(\theta) = L_0 + L_1 \cos(N_r\theta - \pi)$$

$$L_D(\theta) = L_0 + L_1 \cos\left(N_r\theta - \frac{3\pi}{2}\right)$$

Furthermore, mechanical equations can be found from Newton's law of the conservation of energy. Newton's law states that

$$J\frac{dw}{dt} = T_e - T_L - Bw$$

where J is the rotor and load moment of inertia, ω is the rotor velocity in mechanical radians per second, T_e is the torque generated by the motor, T_L is the load torque, and B is the rotor and load viscous friction coefficient. Total electromagnetic torque produced by the motor is, then, the sum of the torques produced by the motor phases

$$T_e = \sum_{x=1}^{m} 0.5\, i_x^2 \frac{dL_x}{d\theta}$$

where m is the phase number, i_x is the winding current in x-phase, and L_x is the inductance function of x-phase winding. Using the law of conservation of energy, the torque generated by i_A for the variable reluctance stepper motor assuming no magnetic saturation is

$$T_A = -\frac{L_1 N_r}{2} \sin(N_r\theta) i_A^2$$

where T_A is the torque generated by the current in A-phase. Summarizing, the differential equations for the three-phase variable reluctance stepper motor are

$$\frac{di_A}{dt} = \frac{1}{L_A} V_A - \frac{R_A i_A}{L_A} + \frac{L_1 N_r}{L_A} i_A w \sin(N_r\theta)$$

$$\frac{di_B}{dt} = \frac{1}{L_B} V_B - \frac{R_B i_B}{L_B} + \frac{L_1 N_r}{L_B} i_B w \sin\left(N_r\theta - \frac{2\pi}{3}\right)$$

$$\frac{di_C}{dt} = \frac{1}{L_C} V_C - \frac{R_C i_C}{L_C} + \frac{L_1 N_r}{L_C} i_C w \sin\left(N_r\theta - \frac{4\pi}{3}\right)$$

$$\frac{dw}{dt} = \frac{T_e}{J} - \frac{T_L}{J} - \frac{Bw}{J}$$

$$\frac{d\theta}{dt} = w$$

$$T_e = -\frac{L_1 N_r}{2}\left[\sin(N_r\theta)\, i_A^2 + \sin\left(N_r\theta - \frac{2\pi}{3}\right) i_B^2 + \sin\left(N_r\theta - \frac{4\pi}{3}\right) i_C^2 \right]$$

Differential equations for the four-phase variable reluctance stepper motor are

$$\frac{di_A}{dt} = \frac{1}{L_A}V_A - \frac{R_A i_A}{L_A} + \frac{L_1 N_r}{L_A} i_A w \sin(N_r \theta)$$

$$\frac{di_B}{dt} = \frac{1}{L_B}V_B - \frac{R_B i_B}{L_B} + \frac{L_1 N_r}{L_B} i_B w \sin\left(N_r \theta - \frac{\pi}{2}\right)$$

$$\frac{di_C}{dt} = \frac{1}{L_C}V_C - \frac{R_C i_C}{L_C} + \frac{L_1 N_r}{L_C} i_C w \sin(N_r \theta - \pi)$$

$$\frac{di_D}{dt} = \frac{1}{L_D}V_D - \frac{R_D i_D}{L_D} + \frac{L_1 N_r}{L_D} i_D w \sin\left(N_r \theta - \frac{3\pi}{4}\right)$$

$$\frac{dw}{dt} = \frac{T_e}{J} - \frac{T_L}{J} - \frac{Bw}{J}$$

$$\frac{d\theta}{dt} = w$$

$$T_e = -\frac{L_1 N_r}{2}\left[\sin(N_r \theta) i_A^2 + \sin\left(N_r \theta - \frac{\pi}{2}\right) i_B^2 + \sin(N_r \theta - \pi) i_C^2 + \sin\left(N_r \theta - \frac{3\pi}{2}\right) i_D^2 \right]$$

The previously given model does not account for magnetic saturation of the ferromagnetic material used to construct the variable reluctance stepper motor. A common and effective way to account for the magnetic saturation is to replace the torque expressions with an expression that is linear instead of a quadratic, in-phase current. For example, the torque due to the current in A-phase is modeled as $T_A = -k_T i_a \sin(N_r \theta)$.

Additionally, for three phases is

$$T_e = -k_T\left[\sin(N_r \theta) i_A^2 + \sin\left(N_r \theta - \frac{2\pi}{2}\right) i_B^2 + \sin\left(N_r \theta - \frac{4\pi}{3}\right) i_C^2 \right]$$

and for four phases is

$$T_e = -k_T\left[\sin(N_r \theta) i_A^2 + \sin\left(N_r \theta - \frac{\pi}{2}\right) i_B^2 + \sin(N_r \theta - \pi) i_C^2 + \sin\left(N_r \theta - \frac{3\pi}{2}\right) i_D^2 \right]$$

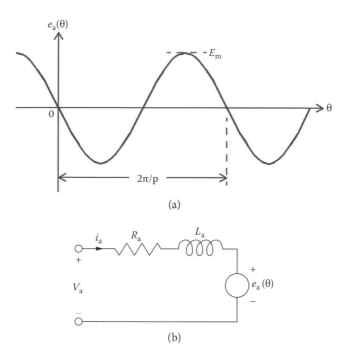

FIGURE 8.35 Equivalent circuit of a permanent magnet or hybrid stepper motor.

where k_T is the torque constant equal to zero speed in one phase on holding torque.

In the same manner, the equivalent circuit for modeling a permanent or hybrid stepper motor is shown in Figure 8.35.

Analyzing the circuit, voltage equations for each phase are,

$$u_a = Ri_a + L\frac{di_a}{dt} + e_a \qquad u_b = Ri_b + L\frac{di_b}{dt} + e_b$$

The voltage source $e_a(\theta)$ represents the motor back electromotive force, which can be approximated as a sinusoidal function of the rotor position, for the a- and b-phases

$$e_a(\theta) = -p\psi_m \sin(p\theta)\frac{d\theta}{dt} = \frac{d\psi_{am}}{dt}$$

$$e_b(\theta) = -p\psi_m \sin\left(p\theta - \frac{\pi}{2}\right)\frac{d\theta}{dt} = \frac{d\psi_{bm}}{dt}$$

where ψ_m is the maximum flux linkage, and $\frac{d\theta}{dt} = w$ is the angular velocity. Thus, the torque generated is

$$T_e = -p\psi_m \left[i_a \sin(p\theta) - i_b \sin\left(p\theta - \frac{\pi}{2}\right) - T_{dm} \sin(2p\theta) \right]$$

$$p = \frac{360}{2m * Step}$$

where T_{dm} is the maximum detent torque, p is the number of pole pairs, m is the phase number, and *step* is step-angle in degrees. Mechanical equations are

$$J\frac{dw}{dt} = T_e - T_L - Bw \qquad \int \frac{dw}{dt} = w \qquad \int w = \theta$$

where J is the rotor and load moment of inertia, ω is the rotor velocity in mechanical radians per second, T_e is the torque generated by the motor, T_L is the load torque, B is the rotor and load viscous friction coefficient, and θ is the position angle.

As an example, these two models are programmed in LabVIEW. Figure 8.36 shows the block diagram of a hybrid stepper motor model, and Figure 8.37 shows the control panel of this stepper motor. As seen in the latter figure, the stepper motor is controlled at the 200 degrees angle position (see "theta" graph). Also, the velocity angle is plotted ("w" graph) and the torque generated ("Te" graph).

SOLIDWORKS AND LABVIEW

In addition, SolidWorks and LabVIEW can be used together to perform the test and validation steps leading into the design process. By using a LabVIEW project and the NI SoftMotion application, scientists and engineers can apply the projects they created with SolidWorks as models in LabVIEW.

This embedding process requires LabVIEW 2009 or later, the LabVIEW NI SoftMotion Module Standard or Premium, SolidWorks 2009, and SolidWorks Motion Simulation. Thus, using NI SoftMotion with SolidWorks can simulate the system with current motion profiles, mechanical dynamics as mass and friction effects, cycle times, and individual component performance. In addition, LabVIEW can easily provide a high-level function block programming language so scientists and

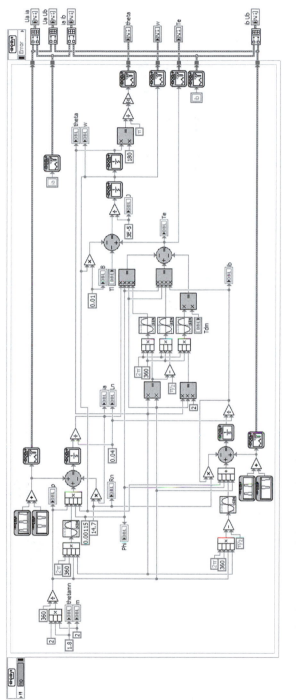

FIGURE 8.36   Block diagram of the variable reluctance stepper motor model in LabVIEW.

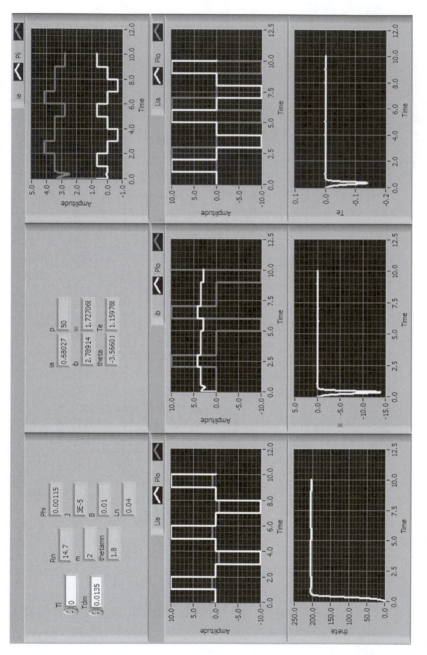

FIGURE 8.37 Control panel of the variable reluctance stepper motor model.

engineers can program a motion control system without the need for control engineering expertise. Some applications related to this embedded system are motion trajectory design; visualization of the 3-D animations on SolidWorks; collision detection between mechanical entities; and time studies for mechanical dynamics, current motion profile, motor management, drivers, and transmission sizing.

The following steps show how to link SolidWorks and LabVIEW. It is assumed that a new LabVIEW project is created:

1. Launch SolidWorks and open the file needed for simulating. Notice that the assembly and motion study must be ready to simulate with all the constraints and motors properly configured.

2. Right-click the **My Computer** item in the LabVIEW **Project Explorer** window and select **New>>SolidWorks Assembly** from the shortcut menu to open the **Import Axes from Assembly File** dialog box.

3. Select the SolidWorks assembly to add to the LabVIEW project.

4. Click **OK**. Then the selected SolidWorks assembly is already added to the **Project Explorer** window, including all motors contained in the SolidWorks motion study. If the SolidWorks assembly contains multiple motion studies, choose the motion study to add to the project using the **Select Motion Study** dialog box.

5. Right-click the SolidWorks assembly in the **Project Explorer** window and select **Properties** from the shortcut menu to open the **Assembly Properties** dialog box. In the **Data Logging Properties** section, specify a name for the log file and place a checkmark in the **Log Data** checkbox. This logs position, velocity, acceleration, and torque data for the simulation to the specified file name in the LabVIEW Measurement (*.lvm) format.

At this point, the SolidWorks assembly is already added to the LabVIEW project, as seen in Figure 8.38a. It shows the motors and sensors previously assigned to SolidWorks. To simulate using the SolidWorks motors included in the model, the motors need to be linked with NI SoftMotion axes. The latter are used when creating motion profiles with the NI

(a)

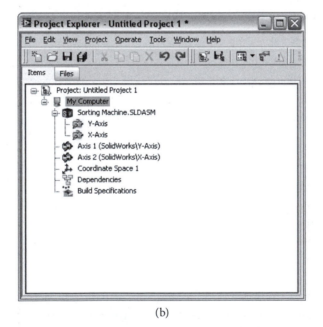

(b)

FIGURE 8.38　SolidWorks assembly added to (a) LabVIEW project, and (b) axes and coordinate space objects in a LabVIEW project.

SoftMotion function blocks. Complete the following steps to add axes to the project:

1. Right-click the **My Computer** item in the **Project Explorer** window and select **New>>NI SoftMotion Axis** from the shortcut menu to open the **Axis Manager** dialog box.

2. Select **Add New Axis.** The new axis automatically binds to an available SolidWorks motor.

NI SoftMotion axes can be grouped into coordinate spaces to perform coordinated moves using multiple axes simultaneously. Figure 8.38b shows axes and coordinate space objects inserted into the LabVIEW project.

Axes associated with SolidWorks motors are assumed to be servo motors because minimal configuration is needed to get started. An axis can be changed through configuration settings:

1. Right-click the axis in the **Project Explorer** window and select **Properties** from the shortcut menu to open the **Axis Configuration** dialog box (Figure 8.39).

2. On the **Axis Setup** page, confirm that the **Axis Enabled** and **Enable Drive on Transition to Active Mode** checkboxes contain checkmarks. This automatically activates all axes when the NI Scan Engine switches to Active mode.

Finally, motion profiles for simulation can be created with the SolidWorks assembly using the NI SoftMotion function blocks on the **NI SoftMotion>>Function Blocks** palette. With these function blocks, scientists and engineers can perform gearing; implement straight-line, arc, and contoured moves; and read status and data information. Figure 8.40 shows the SolidWorks assembly embedded into a VI.

In this way, a solar tracking system was modeled in SolidWorks, as seen in Figures 8.41 and 8.42 and the manufactured prototype in Figure 8.43.

FUTURE PERSPECTIVES

As seen in this chapter, solar energy is an important source of power for generating electricity. This chapter explored several applications of renewable energy technologies, including in-depth examinations of photovoltaic

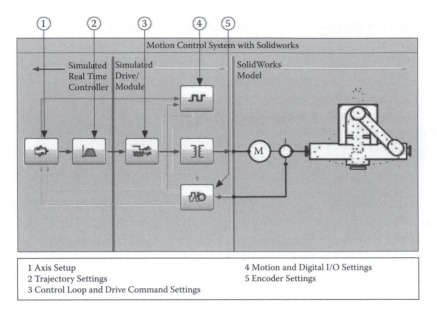

FIGURE 8.39   Axis configuration dialog box.

solar cells, solar thermal generation systems, and mechatronic-oriented solar tracking system modeling and design.

Today, both the industry and academic sectors are concerned about sustainability and developing environmentally friendly technologies. Because of this, new solutions combining photovoltaic cells with organic

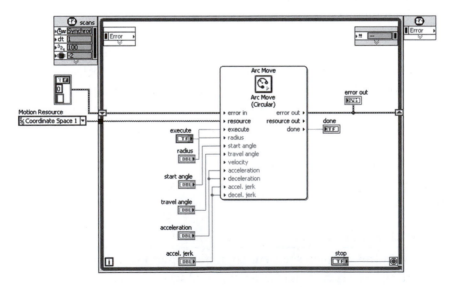

FIGURE 8.40   SolidWorks assembly embedded into the LabVIEW Main VI.

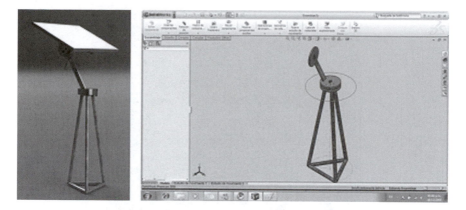

FIGURE 8.41   SolidWorks assembly.

materials are being developed. Additionally, new buildings are incorporating solar technologies to take advantage of the sunlight in the form of building integrated photovoltaics to generate electricity or heating exchangers.

However, solar energy faces challenges such as the amount of space required by solar plants for placing photovoltaic cells to generate electricity. Today, these plants need approximately 17 km² per plant to use solar cells. Research needs to be conducted to increase the efficiency of

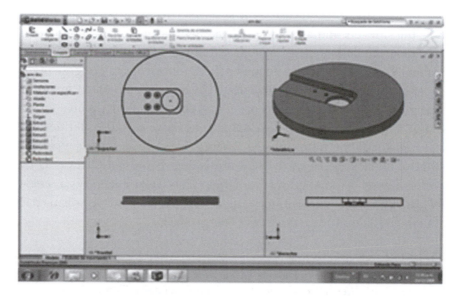

FIGURE 8.42   SolidWorks model.

FIGURE 8.43   Solar tracking system manufactured using SolidWorks design.

photovoltaic cells through concentration and through other methods of increasing overall system efficiency.

Another area of research needed to foster widespread adoption of solar energy is the improvement of energy storage options, such as batteries. Energy must be stored to ensure the levelized effectiveness of solar energy after the sun sets and around the clock. Significant advances in battery technology and other energy storage systems should occur.

Finally, research and development (R&D) investments are growing in the area of solar energy technologies, applications, automated manufacturing and test, and field monitoring and control. Many groups, associations, and researchers are developing innovative new possibilities that will help shape the future of solar energy. Fossil fuels finite supply and negative impact on the earth are diminishing its competitiveness as the world's energy demand grows ever larger. Thus, solar energy, as a renewable energy source, will continue to grow in popularity as a promising part of the solution for the earth's future energy needs.

REFERENCES

Appleyard, D. (2009). *Solar Trackers Facing the Sun*. Renewable Energy World.
Boyle, G. (2004). *Renewable Energy, Power for a Sustainable Future*. Oxford.
Chen, A. (2009). *Solar System Cost Report*. Berkeley Lab.

Goswami, Y., F. Kreith, and J. Kreider. (2000). *Principles of Solar Engineering.* Taylor & Francis.

Markvat, T. (2000). *Solar Electricity.* Wiley & Sons.

National Instruments. (2007). *Getting Started with LabVIEW.* United States of America.

National Instruments. (2009). *LabVIEW 2009 Programming Software.*

National Instruments. (2009). *LabVIEW: The Software that Power Virtual Instrumentation.*

National Instruments. (2010). *Getting Started with NI SoftMotion for SolidWorks.*

National Instruments. (2010). *NI CompactRIO Dual Axis Sun Tracker Example Design.*

National Instruments. (2010). *Solar Power Resource Kit for Automated Test.*

National Instruments. (2010). *Toolkit for I-V Characterization of Photovoltaic Cells.*

National Renewable Energy Laboratory. (2010). *Concentrated Photovoltaic.*

Nelson, J. (2003). *The Physics of Solar Cells.* Imperial College Press.

Ponce, H. (2009). *Intelligent Control System for a Sustainable Portable Greenhouse Using LabVIEW.* Master's thesis. Mexico, ITESM, Campus Ciudad de México.

Ponce, P., H. Ponce, F. Díaz, et al. (2009). *Invernadero Portátil Inteligente.* ITESM, Campus Ciudad de México.

Ponce, P., H. Ponce, and F. Ramírez. (2008). *Intelligent Control Toolkit for LabVIEW.* ITESM, Campus Ciudad de México.

Ponce, P., R. Vargas, S. Castro, et al. (2009). *Sistema Híbrido de Fuentes Alternas de Energía.* ITESM, Campus Ciudad de México.

PV Trackers. (2010). *The PVT 6.0DX.*

SIEMENS. (2010). *Solar Tracking Control, Compact Controller Solution for Solar Tracking.*

Skyline Solar. (2010). *Introduction to High Gain Solar.*

SolidWorks. (2010). *3D CAD Technologies.*

Sustainable Energy Technologies. (2010). *Concentrated Photovoltaics.*

Weisblatt, N. (2006). *Alternative Energy.* Thomson.

Index